WATER BEING WATER

WRIGHT STATE
UNIVERSITY ART
GALLERIES

Published by Wright State University Art Galleries in conjunction with the exhibition *Art/Science*

Wright State University
Art Galleries, Dayton, Ohio
September 6 –
October 16, 2005

Purdue University Galleries
West Lafayette, Indiana
October 24 –
December 4, 2005

Wright State University
Art Galleries
A132 Creative Arts Center
3640 Colonel Glenn Highway
Dayton, Ohio 45435

ISBN 978-0-932706-23-2

Printed by Meridian Printing
East Greenwich, Rhode Island

To Sheryl and Joseph

For their support and trust, thanks to the John Simon Guggenheim Foundation, Bush Foundation, McKnight Foundation and the MacDowell Colony. Thanks also to the Minneapolis College of Art and Design, Wright State University and Ron Geibert.

David Goldes

ACKNOWLEDGEMENTS

Water Being Water and the exhibition *Art/Science* were made possible by the contributions of many individuals. From within the university, generous financial support was provided by Provost David R. Hopkins; Deans Mary Ellen Mazey, James Brandeberry, and Michele Wheatly; and the Department of Art and Art History. Outside support came from Kodak Versamark, Purdue University, the University of Dayton, and Fairborn Camera. WSU Art Galleries Coordinator Tess Cortés and Purdue University Galleries Director Craig Martin and Physics Outreach Coordinator Julie A. Conlon provided strong organizational and programming leadership, and London-based science writer Philip Ball and Minneapolis art critic Vince Leo contributed insightful essays. The exquisite book design was by the talented Jan Jancourt.

During the past two years artist David Goldes joined me in keeping the focus on the unique interdisciplinary nature of this endeavor. It was a pleasure to renew a professional relationship begun 10 years earlier when David participated in my Photography in the 1990s: *fifty portfolios* CD-ROM.

Professor Ron Geibert
Project Director and Curator
Art and Art History
Wright State University

INTRODUCTION

IQ. DNA. MRI. FYI. In an ever-changing world, we find ourselves conversing with abbreviated acronyms and phrases. They drive today's economy, political agenda, and water cooler chitchat. They define our use of resources and the attention given to matters. Who we are, or think we are, are wrapped around their brevity. In our hurried attempt to sort through the complexities of life we use them to only answer the obvious—How? Unfortunately, an equally important question, Why? is often ignored.

The solving of the how of things has generally been left to the scientists and the reasons as to the why for artists to decipher. This publication and exhibition takes a new approach by making the two questions inseparable. Using the simplest of elements, H_2O, a scientist turned artist, and two writers, one from the sciences and the other from the arts, have begun a new conversation. It is a discussion that deserves our attention and participation. -RG

A TABLETOP BEYOND CONTROL IS A WORLD BEYOND DESIRE

VINCE LEO

The photographic tabletop is pure space: plastic, empty, dimensionless, free of the world, divorced from history, replete with immanence. Unlike other photographic spaces, the tabletop is under the total control of a photographer. Everything from camera angle to subject to scale is a completely conscious decision; nothing is left to chance, and no strings attached. The tabletop is also the production site of photographic still life, the purest manifestation of photographic desire; everything we see triangulates with a photographer's intention and interior state and nothing more. Every shadow corresponds to a particular intuition; every highlight a wish fulfilled. On the tabletop, photographic control and a photographer's interior space become a singular agency; decision and desire united in the same set of actions. David Goldes' still lifes are exactly that: products of his control and desire manifest photographically; the world not as he knows it, but as he would like it to be.

Goldes' first and most crucial decision is to choose a creative partner to his photographic self—in his case, an agent so common it escapes notice: water. Even though there are other arguably important materials in his photographs, water is the constant, the tie that binds. It's hard to think of an idée fixe more challenging. The first problem is that, on first glance, water is nothing all that special: Every third grader knows that water covers two-thirds of the earth's surface and makes up over 60 percent of our body mass. The second is that the lesser-known qualities of water—the things that make it unique to our experience—are mostly hidden. Mysteries such as surface tension, wave generation, and state transformation are difficult processes to make visible.

Water is also a near-timeless, self-generating network, an endless cycle through time and space that has eluded specific scientific description. We wash our cars in the water we wash our children in; we irrigate our farms with the water used to hose down civil rights marchers; we drown in the water that we let run over our hands just because it feels good. Every ounce of water commands a narrative so rich as to be almost unimaginable. The water in any one of Goldes' photographs might have been the water from the Johnstown Flood, from an Esther Williams movie, from my daughter's baptism, or from all three.

In addition, water is endlessly metaphorical. From the wine-dark sea to the babbling brook, from the parting of the waves to the water become blood, water embodies important understandings that have no other method of expression. Its sparkling transparency has become a metaphor for the very possibility of life. In the same way, it's difficult to look at a glass of water without thinking of pollution, global warming, and the pernicious effects of human agency on the natural world. In many ways, water has come to stand for both the natural world and the natural world defiled. Like no other substance, water carries with it all the ambiguities we now understand about the dark side of human agency and desire left unchecked.

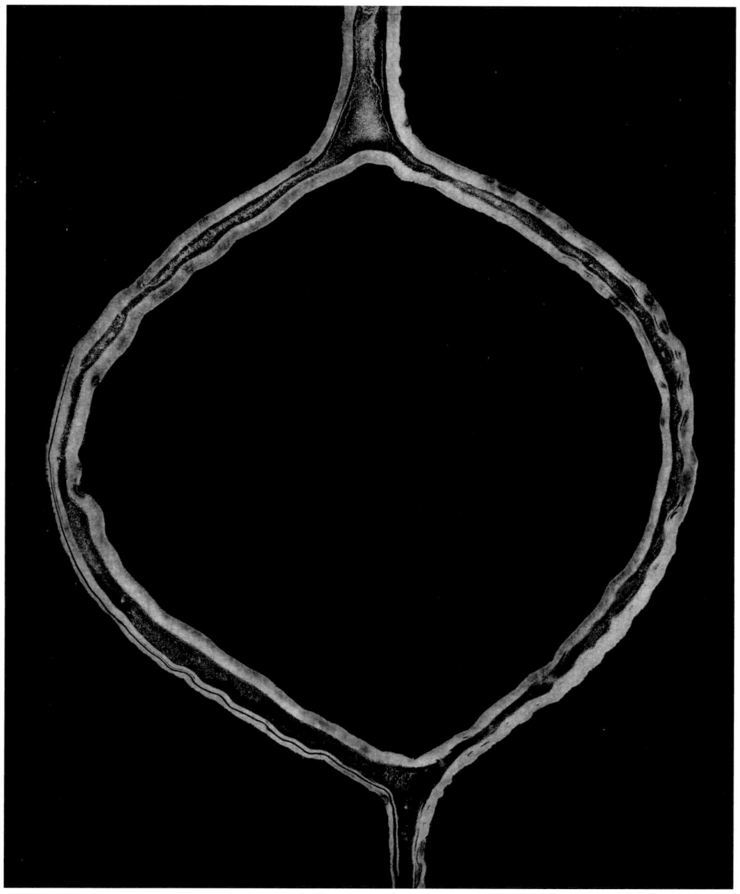

(Fig. 1) PATHS + TRACES #2, 2002
INTAGLIO PRINT

When Goldes chose water, he was choosing the most ubiquitous and mysterious liquid in human experience and the most symbolic substance in human history. He was also choosing one of the most difficult and malleable naturally occurring substances to photograph. Like most materials, the photographic properties of water change with light intensity and direction, but that's just the beginning. Water can be reflective, translucent, transparent, even opaque depending on camera angle, lighting, time factors, etc. Water also refuses to stand still or move in predictable ways, presenting an entirely new set of visual problems for a photographer. Because of all these variables, water forces Goldes to make very specific photographic decisions: stop time or blurred, transparent or reflective, highlights or even surface. On a tabletop, water means an incredible amount of opportunity, and most importantly, a constant level of unpredictability: reflectance can vary within human perceptual error, running water always surprises the camera, and wet surfaces and objects tend to change in unexpected ways. This is not a little thing on a tabletop. Introducing an agency beyond the photographer's control, even in small doses, Goldes changes the equation of decision and desire in a crucial way. When he chooses water, Goldes provides a certain opening for the world to assert itself, to surprise him, to change the picture. Into that opening rushes all the historical signification, chemical characteristics, metaphorical embodiments, and visual properties of water and the world.

Goldes transcribes water's narrative span and visual possibilities with the most abstract, economical, and graceful of photographic languages: the gray scale. From black to white to the infinite shades of tone between, the gray scale is the purest invention of photochemical reproduction. Even though it represents a direct or indexical correlation between light intensity and tonal reproduction, it also gives the photographer the most opportunity to manipulate those correlations, to transform the jarring highlights of the original scene into the smooth midtones of the final print. It's only seldom that we see the kind of creative tonal manipulation that Goldes brings to every photograph he makes. Combining correspondence and manipulation into a seamless descriptive device, Goldes extends physical description into the metaphysical structure of metaphor. There are inky blacks surrounding pins; impenetrable, suggesting both the logic of shadows and the limits of knowledge. There are highlights in gathering pools; immaterial, active, an outline of abstract space and time. There are gray tones in running water, infinite, changing, the everyday in the timeless, describing endless change as a moment of perception. This is not what water looks like; this is what water means. Without us knowing it, Goldes has turned the water in his photographs into something else, something capable of translating water's creative agency into suggestive symbolic language.

Goldes utilizes water's physical properties in another way unique to his still lifes. He doesn't create a scene and then photograph it; instead, he creates a situation that activates water, and then photographs the results or duration of that activation. Goldes' still lifes are like little performances with a cast of two. We'd think of them as science experiments except that they don't have any of the goals associated with scientific experimentation: they don't prove any outstanding theoretical or mathematical proposition, nor do they extend the reach of human knowledge of the natural world. Instead of goals, Goldes' experiments have outcomes: still points that are not only about the natural world, but also about the representational world. Under the guise of scientific experiment, Goldes conducts experiments in metaphor, in the possibility of transforming the odd and beautiful characteristics of water into representations of human experience. Once again, Goldes compromises his control of the tabletop by releasing water's creative agency in the production of these metaphors. Water motivates transparently, hides effortlessly, transforms ceaselessly, sets in motion relentlessly, responds gladly, denies implicitly, surprises without guile, manipulates without conscience; a source of endless possibility and undeniable causality. In Goldes' still lifes, water is what the world visits on our lives and what our lives visit on the world. He weaves the loom, he cuts the thread, he displays the final fabric. Water does the rest.

Within the overarching language of tone and the operational mode of experiment, Goldes' metaphors take different shapes from one body of work to another. The prints created through the collaboration between Cole Rogers and Goldes, *Paths and Traces*, **(Fig. 1)** are made by running water, letting it eddy and pool, and then transcribing the wanderings into stable etchings. The liquid looks fibrous, like traces of biological structure whose paths are both wild and purposeful. As structure and wildness overlap and coalesce, we move from the pictures' myriad interior interactions to their long wandering line and then back again. Representing cellular differentiation and organic unity at the same time, these prints suggest a place in which the limitless and the causal exist in an ongoing harmony. Using the barest of tonalities—basically black field and white line—they hint at Goldes' larger thesis: that control and unpredictability are not a zero-sum game but a matrix for desire, for meaning beyond the immediately foreseeable needs of the present.

Everything expands geologically in the *Flow Pictures*. In this work, Goldes set up his tabletop with glass and back lighting and then covered it with caulk and other substances. Next came a water pump, which redirected water after it flowed through the solid material back up to the tabletop to repeat the process over long periods of time. Goldes photographed the results of water cutting channels though the materials, creating stress lines and patterns. Once again, water is the agent provocateur, the activating force of the work, the substance that literally shapes the outcome. Like all geologies, these miniature riverbeds are discussions of power: the power of deep time and the power of instantaneous catastrophe to reinvent the world. Moving beyond theoretical description, Goldes' work describes the world as a combination of both, but most importantly as a place in which power is constant and unyielding, a dark presence embodied in both the inactive yet resistant mass of solids and the torrential movement of the water. Importantly, the tonal differentiation between resistance and movement is barely discernible. Floating above it all appear certain highlights, a product of water interacting with light. Call it hope or a positive spin or an unexpected manifestation of desire against all the physical evidence, these highlights are the beginning of a new way to think about power: gloomy physical dialectic or not, the metaphysical is always present, an agent of representational transformation and true change. Power beyond control; hope beyond desire.

Goldes' third body of still lifes brings water into the realm of the domestic. Exploring everything from surface tension to vortex to weight displacement, Goldes never gets more complicated than two champagne glasses. Creating situations in which water envisions its miraculous self among the aluminum cans, coffee cups, bowls, and strings, Goldes reintroduces the wonder and vulnerability of the natural world to the order and regularity of domestic desire. Pins float for an instant, a vortex spins confined and beautiful, a homemade turbine powers nothing. The combination of the natural and the domestic on Goldes' tabletop feels a lot like looking into a family scene. The attempts to sustain structure, functionality, and harmony are constantly challenged by the vulnerability, unpredictability, and individual agency of human beings. The difference is that on Goldes' tabletop, it all works out. The miraculous and the ordered don't simply exist side by side, they activate each other, make each other visible. Once again, this is not only the domestic of David Goldes' desire and decisions, it is a domestic beyond desire and decision, a domestic in which the miraculous in the form of water is free to unify and liberate.

The larger metaphor in all of Goldes' still lifes is in his method: in the way he tempers complete photographic control with the immanent agency of water. Water undermines control, and in the process expands desire beyond Goldes' conscious purpose. Being able to realize pure desire through complete control is not only the proposition guiding the photographic tabletop, it's also a specter haunting postmodern Western culture. If modernism gloried in the utopian possibilities of control and the ultimate good of personal desire, postmodern society is saddled with the morning after: failed architectures, depleted ecosystems, meaningless and destructive economies. Rather than accept this set of propositions, Goldes has taken a courageous step toward something new. His genius is to set up situations in which there is a delicate balance between his decisions, his desires, and the vagaries of the world. That balance is where the unpredictable new is manifest, beyond control but not beyond creative agency. In Goldes' work, desire and control are negotiated, not pure; they are discovered, not premeditated. Control is something we seek to undermine through trial and error and hard work; desire is something we noticed on the way to work and realize is gone the next instant. Desire and decision, linked through discovery, is also the way Goldes repairs the world, makes it new again as a place of beauty and wonder, understood for its possibilities rather than its limitations. This is the essence and the blessing of David Goldes' tabletop: beyond control, beyond desire, a new world is always appearing.

THE LIFE OF WATER

PHILIP BALL

Water does strange things to people. It makes them dream, even to death: water entranced poor Narcissus. We smell the bromide tang of the ocean and dream of voyages and exploration (look what happened once we smelled water on Mars). The babble of the stream, the 'music of humanity' as William Wordsworth called it, makes us dream of serenity and solitude. We dream that water is a universal panacea, that it will salve our wounds: 'We attribute to water virtues that are antithetic to the ills of a sick person', says French philosopher Gaston Bachelard. 'Man projects his desire to be cured and dreams of a compassionate substance.'

People have found remarkable things in that liquid mirror. They have wished to turn water into a fuel, a bountiful source of energy. To the theosophist Theodor Schwenk, flowing water was a sensory organ through which celestial influences enter into the world, receiving 'formative impulses from the spiritual world.' Blessed by the priest, water gave protection from the devil. Sacred water is beyond corruption: nothing can defile it. 'The faithful Persian', claimed the English anthropologist Edward Tylor in 1903, 'may be seen by the side of the little tank where scores of people have been in before him, obliged to clear with his hand a space in the foul scum on the water, before he plunges in to obtain ceremonial purity.'

Scientists are, of course, supposed to dispel such mysticism. But so far they've not been very successful; in fact, they have often succumbed themselves. American biochemist Lawrence Henderson argued in 1913 that water is uniquely 'fit' for sustaining life, in Darwin's sense: its properties seem to be fine-tuned – 'adapted' – to the needs of living organisms. 'Water, of its very nature', he said, 'as it occurs automatically in the process of cosmic evolution, is fit, with a fitness no less marvelous and varied than that fitness of the organism which has been won by the process of adaptation in the course of organic evolution.' From there it is a short step to a theological 'argument from design'. Henderson was aware that his story was veering into this metaphysical realm, and he held back from making any kind of direct statement about whether water is indeed 'purposeful' and speaks of a supreme Designer. But such thoughts were clearly in his mind.

More recently, water science has acquired a rather disreputable image, thanks to recurring claims that the liquid has near-miraculous properties. In the late 1960s, Soviet scientists awakened a frenzy of research which turned into a scandal when they proposed that water can exist in a highly viscous form called 'polywater'. Some researchers worried that a seed of polywater might convert all the world's oceans to this gum, in the same way that they are frozen into a novel form of high-temperature ice, called ice-nine, in Kurt Vonnegut's Cat's Cradle (1963). Yet polywater turned out to be a figment of bad experimental technique. And in 1989, French biologists asserted that water has a 'memory' which enables it to retain the imprint of biological molecules, even after a solution has been so diluted that in chemical terms there is nothing left but pure water. An explanation for homeopathy? The experiments have never been convincingly repeated, and now 'the memory of water' has become a stick with which to beat those scientists foolhardy enough to dip their toes into water.

But that might have been predicted all along, because water can never be a neutral object of study. Scientists may call it H_2O, a humble chemical compound in which the molecules are each made up of two hydrogen atoms and one of oxygen. But we all know that in experiential terms water remains elemental, just as the Greek philosopher Empedocles asserted in the fifth century AD. For Empedocles, the chemist's Periodic Table had only four spaces, and they were filled by earth, air, fire and water. This list of elements might be (it certainly is) wrong in chemical terms, but it corresponds with our intuitions about matter, which has the solidity of earth, the fluidity of water, the flightiness of air and the flickering brightness of fire. Water is an ancient, mythical substance, it is the medium of both life and death, and it challenges profoundly our ability to regard the material world through objective eyes.

THE POETICS OF WATER / Bachelard recognizes many 'elemental', material themes in poetry and literature. 'When I began meditating on the concept of the beauty of matter', he says, 'I was immediately struck by the neglect of the material cause in aesthetic philosophy... In fact, I believe it is possible to establish in the realm of the imagination, a law of the four elements which classifies various kinds of material imagination, by their connections with fire, air, water, or earth... A material element must provide its own substance, its particular rules and poetics.'

Shakespeare is quite explicit about this: his Sonnets 44 and 45 are paeans to the classical elements: slow, ponderous earth and water that conspire to separate him from his lover, and 'slight air and purging fire', 'the first my thought, the other my desire'. But Bachelard has dived deep into the poetics of water (and of fire too), and he identifies several fluid archetypes that describe our relationship with this element. Narcissus illustrates the potency of 'reverie before a watery reflection': 'The reflected world... requires only inaction, only a dreamer's attitude.' It is in this mirror, Bachelard claims, that 'Generalized narcissism transforms all beings into flowers, and it gives all flowers consciousness of their beauty': hence Percy Bysshe Shelley's 'yellow flowers' that 'forever gaze on their own drooping eyes.'

The graceful swan, says Bachelard, represents a kind of 'sanctioned nudity', an image of 'amorous water' revealed in the legend of Leda. Edgar Allen Poe makes much use of the image of heavy, dense water, a liquid that stands for blood, 'our human water loaded with virtue and spirit' as poet Paul Claudel put it. Poe revels too in 'violent water', the wild maelstrom of the oceans that destroys and sucks us under. We know this image well enough: it is the tsunami, the Great Wave in Hokusai's famous woodblock print, the wave that is thought to have swept out of the Aegean Sea in the fifteenth century BC to inundate northern Crete and trigger the collapse of the Minoan civilization. This angry water allegedly buried Atlantis; undoubtedly it has killed people in the thousands along the coasts of Java and Japan.

Violent water is found not only in the ocean; it is also the storm, it comprises the raging cataracts in King Lear that 'drench'd the steeples and drown'd the cocks'. 'There can be no epic without a storm scene', Bachelard says. And water has long been associated with death, so that the dead are considered to set out on a voyage over the seas, into the unknown. 'O death, ancient captain, the time has come!' wrote Baudelaire. 'Let us weigh anchor!' But sometimes death comes back, on board the Marie Celeste. And if not the sea, then death's journey crosses a river: 'all souls', Bachelard claims, 'whatever the nature of their funerals be, must board Charon's boat.'

LIFE'S MATRIX / On the other hand, water is life too. That's no contradiction: water has a multiplicity of roles in our myths and cultures. Schwenk's vortexed, spiritually receptive water is 'living water'; this stuff, apparently being whisked up in David Goldes' Jar (1998), is today being marketed commercially as a 'health-giving' form of water with a vitalized molecular structure. Believe that if you will; but even some scientists speculate that water inside our cells is somehow different from normal water, that it is 'tamed' by its organic environment. The jury is still out on this idea, yet the image itself unconsciously accedes to the poetic image of maternal water, of water as an elixir: in fact, of water as milk. 'The waters, which are our mothers', says a Vedic hymn, 'come to us following their paths and distribute their milk to us.'

Does this image of fecundity play a part in NASA's determination to look for life on other planets by 'following the water'? It is a reasonable principle, at any rate; for we know of no living organism on Earth that can function without water, and likewise, wherever there is water, life seems to seek it out. So this association between water and life commonly passes unchallenged. Even if other liquids might serve as exotic solvents on distant worlds – liquid ammonia or methane in cold places, molten rock in hot ones – it remains exceedingly hard to imagine what form such life might take, and scientists often seems almost embarrassed at even contemplating the idea of a non-aqueous biology.

AN ACTIVE MEDIUM / It's strange to find that, given this article of faith, this near-reverence for water's role in life, biologists have nonetheless tended to relegate it to the status of a mere solvent, a lubricant for life's chemistry, a passive backdrop on which the molecular components of life do their business. It's become increasingly clear that this isn't the case at all. Water is very much an active player in life's chemistry, and biology has evolved to make creative use of some of water's unusual properties.

At the microscopic level, the fundamental unit of life is the single cell. We are each a vast mass of such cells, each of them specialized for a particular function and each collaborating with the others to create an autonomous, integrated being. So to talk about water's fundamental role in biology is to ask: what does it do inside cells?

It fills them, for a start. If the well-known fact that we are mostly (about two thirds) water does not seem to turn us into sloshy water-skins, that is because much of this water is packaged away in cells, swelling them like tiny balloons. Close up, we are like pomegranates. This rigidity that water gives to cells accounts for much of the stiffness of plant tissues and green stems. When plants are deprived of this water, they droop.

Dissolved in the water of our cells are all manner of substances: salts, sugars, proteins, DNA, hormones. Yet cells are not just bags of liquid: they are highly organized, with special compartments where different processes happen – energy production, protein synthesis, chromosome replication and so on. Cells are like intricately orchestrated societies, and most of that orchestration is conducted by proteins, the molecular workhorses of biology.

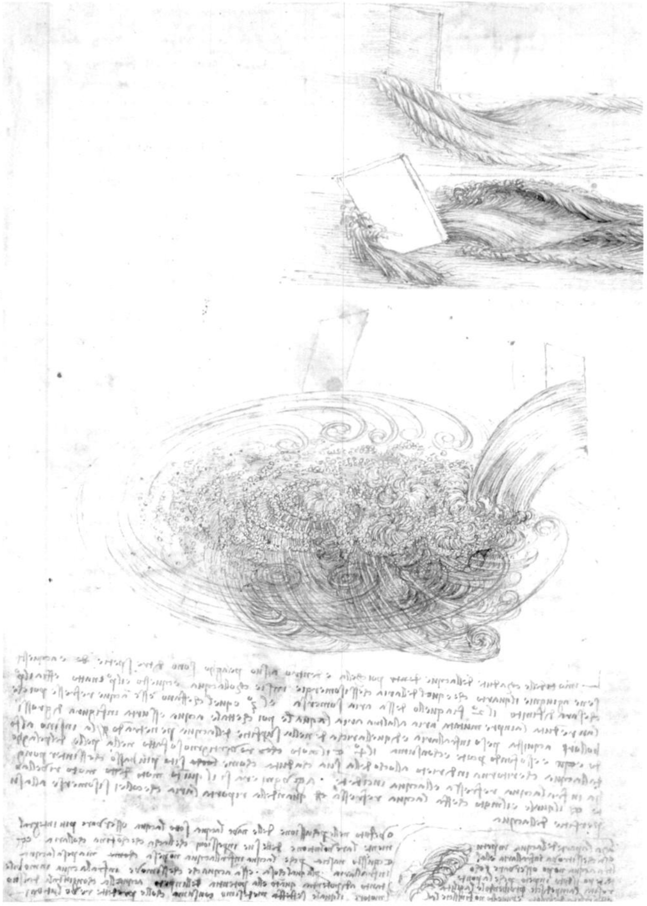

(Fig. 2) STUDIES OF WATER, LEONARDO DA VINCI

Water is what enables proteins to function properly. To begin with, proteins can't find their proper shape unless they are immersed in water. Enzyme proteins are basically long strings of amino acids, like beads on a necklace; but in order to do their jobs, these strings have to be folded up into compact bundles. The forces acting between the amino acids and the surrounding water molecules help to direct this folding process towards the right shape. In other words, water sculpts proteins. And typically, enzymes have to retain some flexibility if they are to function correctly – they mustn't be folded too tightly or rigidly. Water acts as a lubricant to keep them 'soft': if you take them out of water, some enzymes will still work (provided that they are already folded correctly) – but they tend not to work so well, because they become rather too stiff. And even then, the enzymes make sure to bring some tightly-bound water with them.

What is more, water molecules sometimes act as 'snap-on' tools that enable proteins to grab hold of other molecules, or to interact with one another. Some proteins make use of the way chains of water molecules can pass hydrogen ions very rapidly along the chain, like tiny wires conducting electricity. Water molecules are truly nature's little helpers.

THE WET UNIVERSE / In choosing to look for extraterrestrial life where there is liquid water, NASA is arguably being conservative – but it is surely a good place to start. We don't know that life without water is impossible, but we do know that water makes life possible, perhaps even probable. So where should the 'astrobiologists' be looking?

The first thing to understand is that water is not rare in the universe. Hydrogen, whose name means 'water-former', is the most abundant element – around three quarters of all the ordinary matter in the cosmos is hydrogen. (There is a lot of extraordinary, so-called dark matter too, but we can detect that only indirectly, and have no idea what it is made of.) All this hydrogen is the legacy of the Big Bang, for in material terms that primal event of creation gave us rather little else. Oxygen came later, forged inside stars by the repeated uniting of hydrogen atoms into successively bigger atoms, heavier elements: a process of nuclear fusion, the origin of starlight.

Stars are good oxygen factories, and have made it the universe's third most abundant element. When hydrogen and oxygen come together, they are rather eager to react to form water, H_2O – that is why the old hydrogen-filled airships burnt so terribly well. This reaction happens at a more leisurely pace in the imponderable expanses between the stars, in the tenuous haze of gas called 'molecular clouds', where water is common. There in the making is the river Nile, the Arabian Sea, the clouds and snowflakes, the juice of cells, the ice plains of Neptune, and who knows what other rivers, oceans, and raindrops on worlds we may never see. For molecular clouds are the birthplaces of stars, which coalesce from dense blobs that collapse under their own gravity. And we now know that many, perhaps most, stars acquire their own planetary systems like our own solar system. They are sure to have water in them.

But if planets are too far from their sun, any water they contain will freeze to ice. This is the case on the outer planets of our solar system: Pluto is more or less a ball of ice, frozen so cold that it is as hard as granite. Uranus and Neptune have thick layers of ice below their outer atmospheres of hydrogen and helium. Saturn's moons are covered in ice, and so are three of the four major moons of Jupiter. There's no water shortage in the solar system – but it's mostly frozen.

The Earth orbits the Sun within a narrow band called the habitable zone. Here the temperature is warm enough for water to exist as a liquid on the planet's surface (that's helped by the presence of natural greenhouse gases in the Earth's atmosphere, which warm up the planet by about 35°C), but not so hot that the water all boils away. This happened on Venus, where the surface temperature is around 500°C. Vaporized into the atmosphere, Venus's water was split by sunlight into its constituent atoms, and the lightweight hydrogen atoms then slipped away from the planet's gravity. When it was newly formed, Venus probably had about as much water as the Earth, but now it has mostly leaked out into space.

And what, then, of Earth's nearest neighbour, Mars? If there is life elsewhere in our solar system, this – as science-fiction enthusiasts know well – is the most likely place to find it. Indeed, until the 1950s it was virtually taken for granted that there was plenty of life on Mars: the seasonal darkening of parts of the planet's surface (now known to be caused by dust storms) was assumed to be due to the growth of plant life. In the 1960s, however, space missions to Mars sent back a disheartening message: the planet seemed to be very cold (between minus 60 and minus 120°C) and bone dry.

But then we looked closer, and found another story. Mars has rivers. At least, that is what they look like, although they too are dry. The Mariner 9 spacecraft in 1972 showed that the rocky surface is laced with channels that look just like river deltas on Earth. There is no known geological process, except for erosion by running water, that can create such features. It now seems likely that the surface of Mars did indeed once have rivers, perhaps even lakes and oceans. But that was billions of years ago – the surface has since dried up and turned frigid. The latest results from spacecraft orbiting the red planet, however, show that there is plenty of water still below the surface. Perhaps it is frozen, like the permafrost of our terrestrial tundra. Maybe there are even subsurface pockets of liquid water, in which case... who knows? In 2004, scientists reported that they had detected formaldehyde and methane in the Martian atmosphere – in tiny amounts, it is true, but nevertheless too much of it to be able to explain as a product of purely geological processes. The alternative – speculative, tantalizing, enthralling – is that this gas could be produced by martian microorganisms, still thriving somewhere down there around remaining pockets of water.

LEONARDO'S VISION / The rivers of Mars are a reminder of the awesome force of flowing water. Some of Mars's dry channels dwarf the Grand Canyon, and it's been claimed that a gully called the Ma'adim Vallis, over a mile deep and 550 miles long, was carved in a matter of months by a flood large enough to drown the states of Texas and California.

The flow of water has always presented a challenge to artists. How do you capture something so dynamic and ever-changing in a static picture? Rather different traditions have arisen in the West and the East.

Western artists have tended to show flow as a play of light on the water surface; traditionally they have painted choppy seas as serried, foam-flecked waves. But Chinese and Japanese artists don't strive for this kind of photographic snapshot realism. Instead, they attempt to portray the essence of flow, or what the 12th-century Chinese critic Tung Yü called 'the fundamental nature of water', schematising it as a series of lines. This can be regarded as an expression of the Taoist conviction that there exists a fundamental simplicity beyond the superficial shapes and forms of the world; to paint a frozen instant would be to paint an illusion. This simplicity is not static; unlike Plato's notion of crystalline ideal forms, the Tao is alive with spontaneity. What could be more representative of the Tao than the currents of a river swirling around rocks? It is precisely this spontaneity that the Chinese classical artist would try to capture with movements of the brush: 'He who uses his mind and moves his brush without being conscious of painting touches the secret of the art of painting', said the writer Chang Yen-yüan in the ninth century.

Curiously, this mode of representing water is very close to the way scientists do it. Those who study fluid flow, called fluid dynamicists, generally show the structure of a flow using so-called streamlines, which, loosely speaking, depict the trajectories that a floating object would follow. Water in Far Eastern art is basically a diagram of streamlines.

There have been attempts to illustrate flow this way in the West, most famously in da Vinci's studies of water **(Fig. 2)**. Leonardo, like the Chinese painters, strove to find the essential forms and patterns of flow beneath the ephemeral splashes and gleams. But unlike them, his interest was not purely aesthetic or philosophical: Leonardo was also a hydraulic engineer, forever devising water-powered machines and scheming to manipulate the flow of rivers. He investigated Archimedes' spirals for lifting water, as well as suction pumps and water wheels. In collaboration with Niccolò Machiavelli, he drew up plans to redirect the flow of the Arno river northwards, away from Pisa, thereby depriving it of its water supply and delivering it into the hands of the Florentines. Leonardo's plans were entrusted to an engineer named Colombino, who botched them, forcing the project to be abandoned.

Leonardo did not become fascinated by water because of his engineering activities; rather, according to art historian A. E. Popham, the latter were a symptom of the former. 'Something in the movement of water, its swirls and eddies, corresponded to some deep-seated twist in his nature', Popham says. Leonardo's self-portrait from 1512 shows his long hair and beard flowing with those same eddies. There was nothing subconscious about this: Leonardo himself said,

> *Observe the motion of the surface of the water which resembles that of hair, which has two motions, of which one depends on the weight of the hair, the other on the direction of the curls; thus the water forms eddying whirlpools, one point of which is due to the impetus of the original current and the other to the incidental motion and return flow.*

This same connection between hair and water is evident in the whorls of Art Nouveau, and can be seen in the water sprites and undines of the English illustrator Arthur Rackham.

Leonardo had recurrent dreams of catastrophic inundation – one

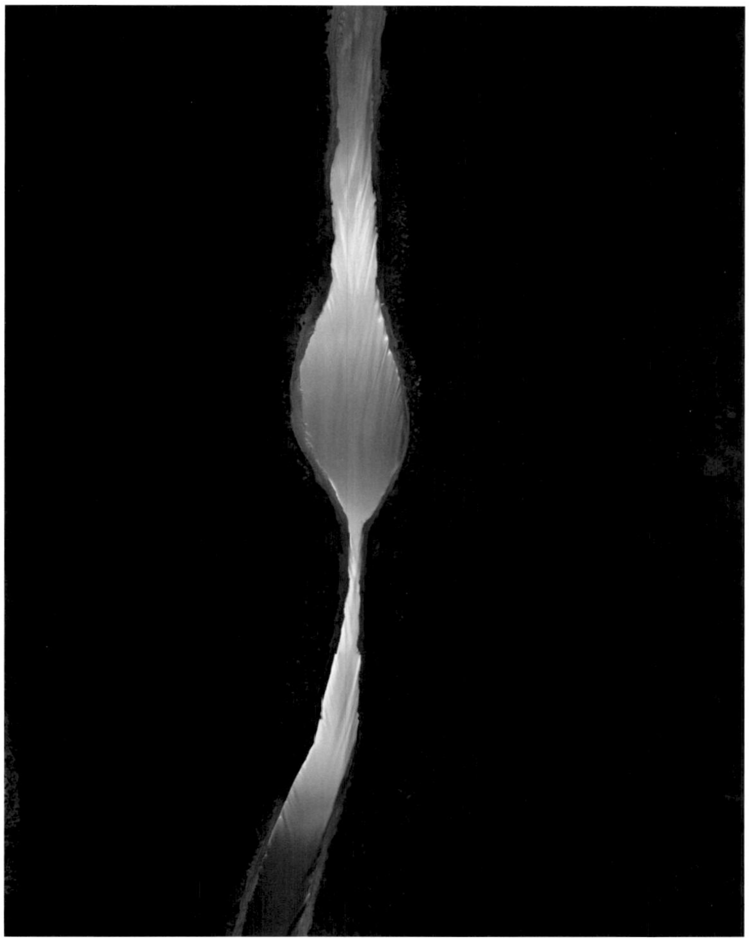

(Fig. 3) FLOW #3, 1998, GELATIN SILVER PRINT

of the clearest artistic manifestations of Bachelard's archetype of violent water. Leonardo made several sketches of monstrous, destructive waves that engulf landscapes and topple cities, where again the flow forms of water are depicted with an attention to detail that is almost gleeful and certainly terrible. 'Among irremediable and destructive terrors', he once wrote,

> *the inundations caused by rivers in flood should certainly be set before every other dreadful and terrifying movement, nor is it, as some have thought, surpassed by destruction by fire... in what terms am I to describe the abominable and awful evils against which no human resource avails? Which lay waste the high mountains with their swelling and exalted waves, cast down the strongest banks, tear up the deep-rooted trees, and with ravening waves laden with mud from crossing the ploughed fields carry with them the unendurable labours of the wretched weary tillers of the soil, leaving the valleys bare and mean by reason of the poverty which is left there.*

How readily Leonardo would have understood the awesome atmospheric eddies of hurricanes as seen by today's weather satellites. Moreover, these gigantic flow forms illustrate one of the fundamental truths about flow that Leonardo was seeking to grasp: within all the turbulent chaos of rushing water, there can arise distinct, robust structures and patterns. Edgar Allen Poe took us into the very eye of one of these 'coherent' flow forms, whipped up in the ocean:

> *Never shall I forget the sensation of awe, horror, and admiration with which I gazed about me. The boat appeared to be hanging, as if by magic, midway down, upon the interior surface of a funnel vast in circumference, prodigious in depth, and whose perfectly smooth sides might have been mistaken for ebony, but for the bewildering rapidity with which they spun around, and for the gleaming and ghastly radiance they shot forth, as the rays of the full moon, from that circular rift amid the clouds which I have already described, streamed in a flood of golden glory along the black walls, and far away down into the innermost recesses of the abyss.*

GO WITH THE FLOW / But these forms are beautiful too. As water flows slowly around an obstacle, a pillar or a branch standing upright in the stream, little vortices are shed from out of the wake, first on one side and then on another, creating a baroque string of whorls called a Kármán vortex street, after the Hungarian physicist Theodore von Kármán. Real streams are flowing too fast to set up such a regular pattern – the wake breaks up into disorganized turbulence – but Kármán vortex streets have been seen in satellite pictures of clouds shaped by air flow, for example as the air streams past a mountain or an island. Similarly, two fluid streams moving past one another in opposite directions can set up regularly spaced eddies like breaking waves at their interface, a phenomenon called a Kelvin-Helmholtz instability. This too can be seen in cloud formations – I've seen them myself, rippling over London's dirty haze.

In nature, however, water generally flows too fast for such regularities – even the broadest, most sluggish rivers are turbulent. Describing turbulence is one of the most daunting challenges for scientists, and no one has yet managed to come up with a completely convincing theory for it. But that needn't worry artists. The British photographer Susan Derges has captured the complex, ephemeral beauty of these turbulent flows in streams and rivers. She immersed huge sheets of photographic paper protected between glass plates just beneath the water surface of the River Taw in Devon, southwest England, and illuminated them at night with a very brief flash of light. This reveals all the little peaks and troughs of surface waves as a kind of shadowgraph. Derges has herself studied Japanese art – she lived in Japan in the 1980s, where she was influenced by the works of Hiroshige and Hokusai – and she is familiar with the Taoist notion of distilling the universal from the particular.

In David Goldes' images of water moving under pressure through clay channels (Flow Series), the flow is both less frenetic and less free **(Fig. 3)**. The streams bulge into aneurysms or branch like arteries. There we see a reflection of the shapes of rivers on a larger scale, as water carves its way through mountain ranges or furrows across broad, sandy plains. These branching forms are one of nature's universal patterns, a common motif of flow in the face of resistance: lightning makes such forks as it finds its way through poorly conductive air, and cracks in brittle materials have a similarly arboreal profile.

WANDERING WATER / Leonardo was a close observer of landscape in general, but natural waters perhaps did most to draw out the artist in the engineer: he spoke passionately of 'beautiful and delightful places with limpid waters through which the green bed of the stream can be seen and the play of the waves rolling through meadows and over pebbles, mingling with blades of grass and with playful fishes and similar subtle detail.' Some of his landscape sketches look astonishingly like satellite photographs, with rivers and tributaries carving mountain ranges into fractal fronds. The resemblance to the venous system of the human anatomy was not lost on this proto-scientist who claimed personally to have dissected 'more than ten human bodies'. True to the Renaissance belief that the 'microcosm' of man reflected the 'macrocosm' of the world, Leonardo considered river waters to be the 'blood of the earth' and claimed that 'the ocean fills the body of the earth with an infinite number of veins of water.'

The river Arno that wanders through the landscape behind Leonardo's Mona Lisa was a favourite icon for Florentine artists – it appears in several works by Antonio Pollaiuolo, for example. Such meanders – one of the characteristics, incidentally, that lend weight to the idea that the channels of Mars are the work of flowing water – were surely a warning to Leonardo the hydraulicist that rivers are not idly to be tampered with. They have a will of their own, which it is not easy to direct down any man-made course. Meanders, named after the famously wandering River Menderes in modern Turkey, seek their own characteristic wavelength – they are one of the river's spontaneous patterns. But they do not stay still. The winding Mississippi can shift its curves by up to 20 metres a year, playing havoc with farmland on the banks.

A river is a dynamic thing, a process more than a feature, a balance between erosion and deposition of silt. This writhing of a river explains how a narrow braid of water can carve out for itself a broad flood plain on a river valley. With each flood, the river renews and moistens the sediments of the plain, producing fertile lands like those of the Nile, Tigris and Euphrates where civilization was born. When the Egyptians painted the river with their famous blue pigment, they knew they were painting the source of life itself.

CREATIVE TENSION / Science, like art, is good at subverting expectations. A substance's properties are a matter of scale; to the water strider and the fishing spider, water has a sturdy skin that can be punctured only with effort, while the Olympic swimmer sees no such barrier obstructing the depths. Surface tension arises because surfaces are expensive: it costs energy to make them, and so a fluid will adapt to minimize the area of its surface. That is what smoothed Narcissus' placid mirror, and it is what pulls raindrops into compact spheres. It is why dew beads up to transform an early-morning spider's web into a lattice woven from strings of tiny pearls.

But if you lower water's surface tension with a film of soap, reducing the energy cost where water meets air, then the liquid can be pulled into the most delicate of membranes, marbled by the reflection of light from the superficial layers of soap. David Goldes explores this exquisite balance between dynamic flow and static force: bubbles sit on wire frames like alchemical glass retorts, they balloon in the breeze, stretched by the air and yet still smoothed by the exigencies of surface tension, for even bubbles do not permit superfluous surface.

These photographs imply a permanence, a structure poised forever in harmonious balance; but that can't be so. Inside the thin film, water flows downwards under the tug of gravity, thinning the skin until it bursts. This rupture is a chromatic epiphany, for the changing thickness of the film picks out different wavelengths of reflected light and the soap film crosses the rainbow before turning silvery and then finally an ominous, fatal black. Goldes' images remind us that, with water, there can be no permanence, but at best only moments of precarious, breathless equilibrium.

Bibliography/Further reading

G. Bachelard (1983). Water and Dreams (Pegasus Foundation).
G. Bachelard (1987). The Psychoanalysis of Fire (Quartet).
P. Ball (1999). Life's Matrix: A Biography of Water (Farrar, Straus & Giroux).
P. Ball (1998). The Self-Made Tapestry: Pattern Formation in Nature (Oxford University Press).
L. Henderson (1958). The Fitness of the Environment (Beacon Press).
E. A. Poe (1988). Tales of Mystery and Imagination (Gramercy). [This version is illustrated by Arthur Rackham.]
A. E. Popham (ed.) (1946). The Drawings of Leonardo da Vinci (Jonathan Cape).

What is it that water does do. -Gertude Stein

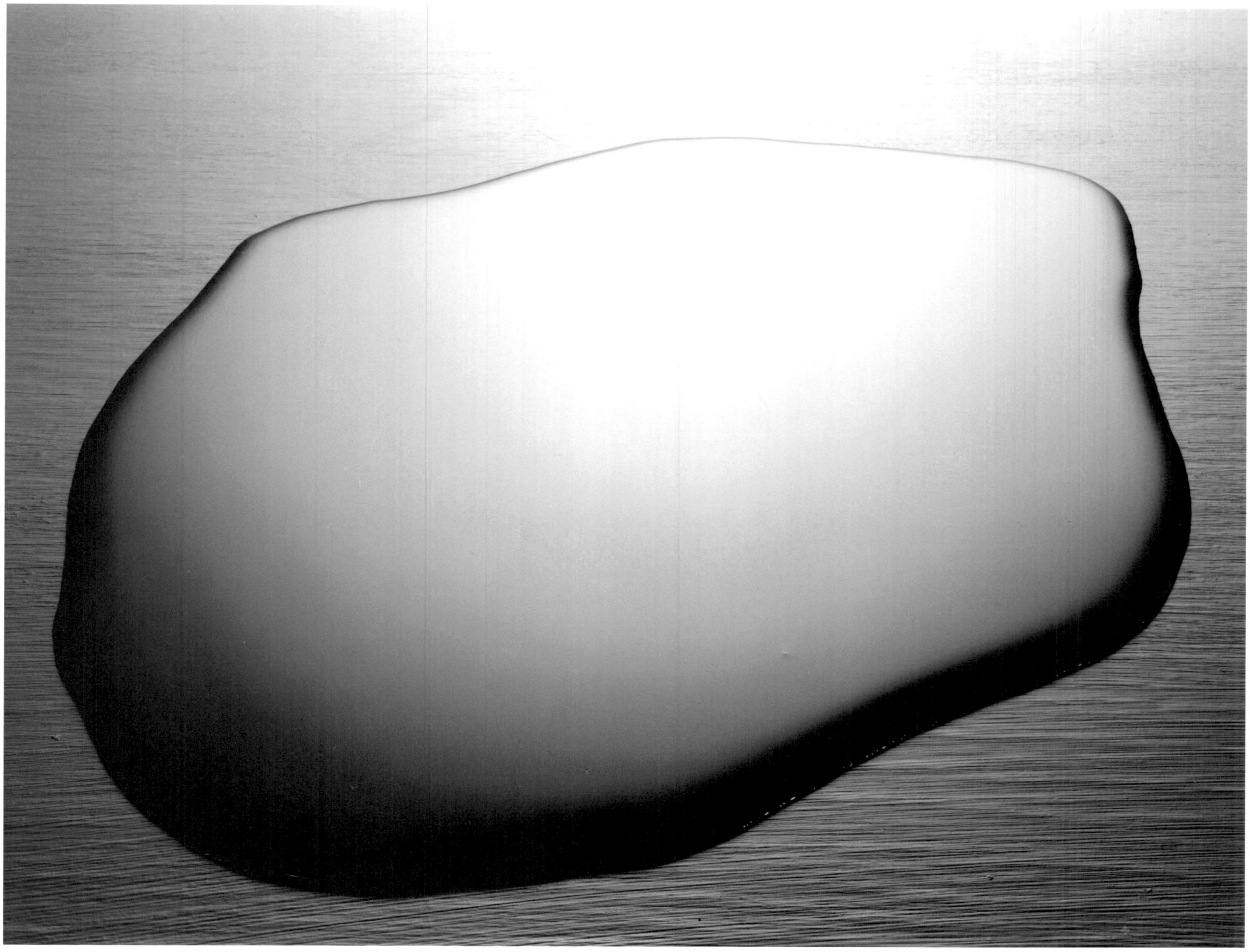

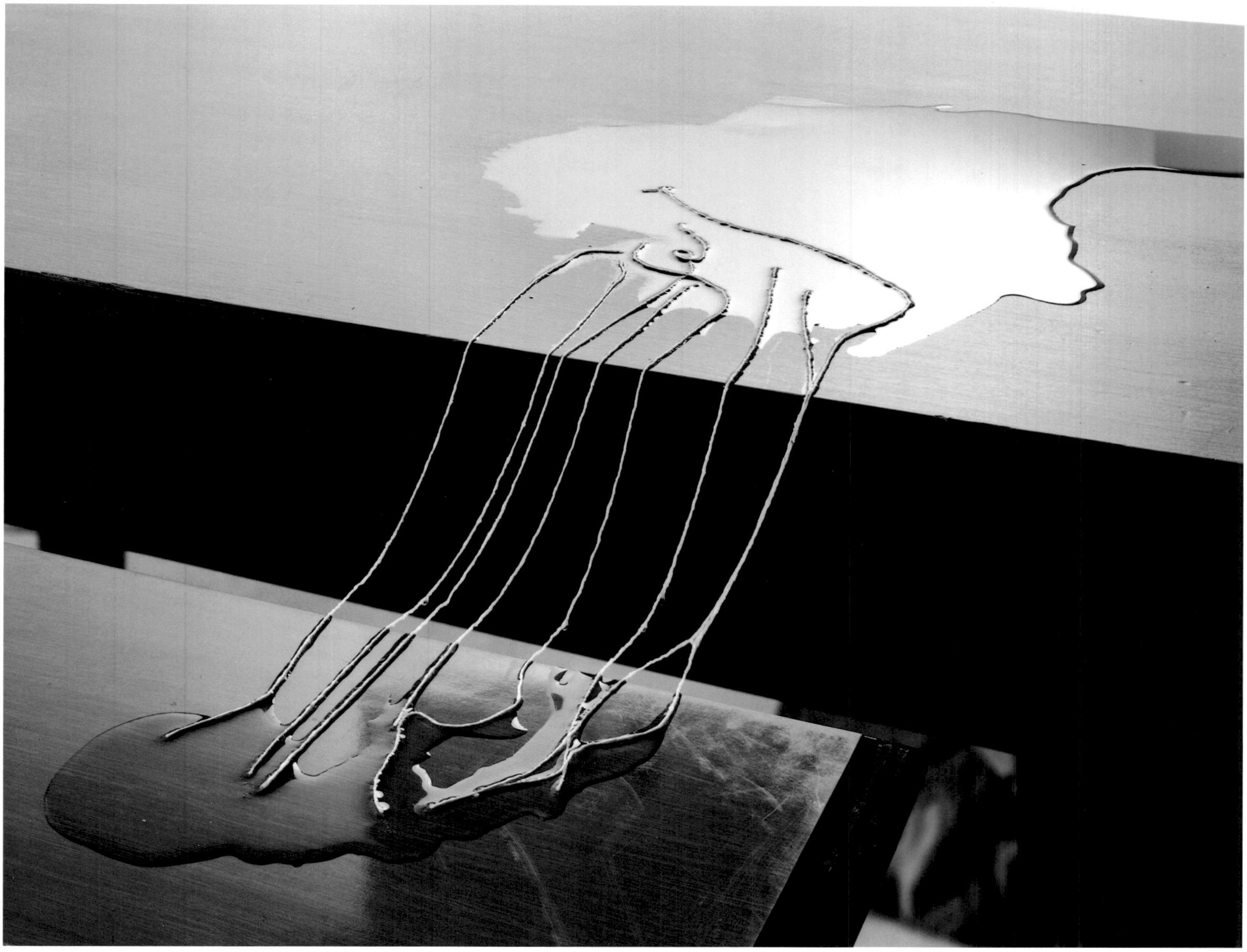

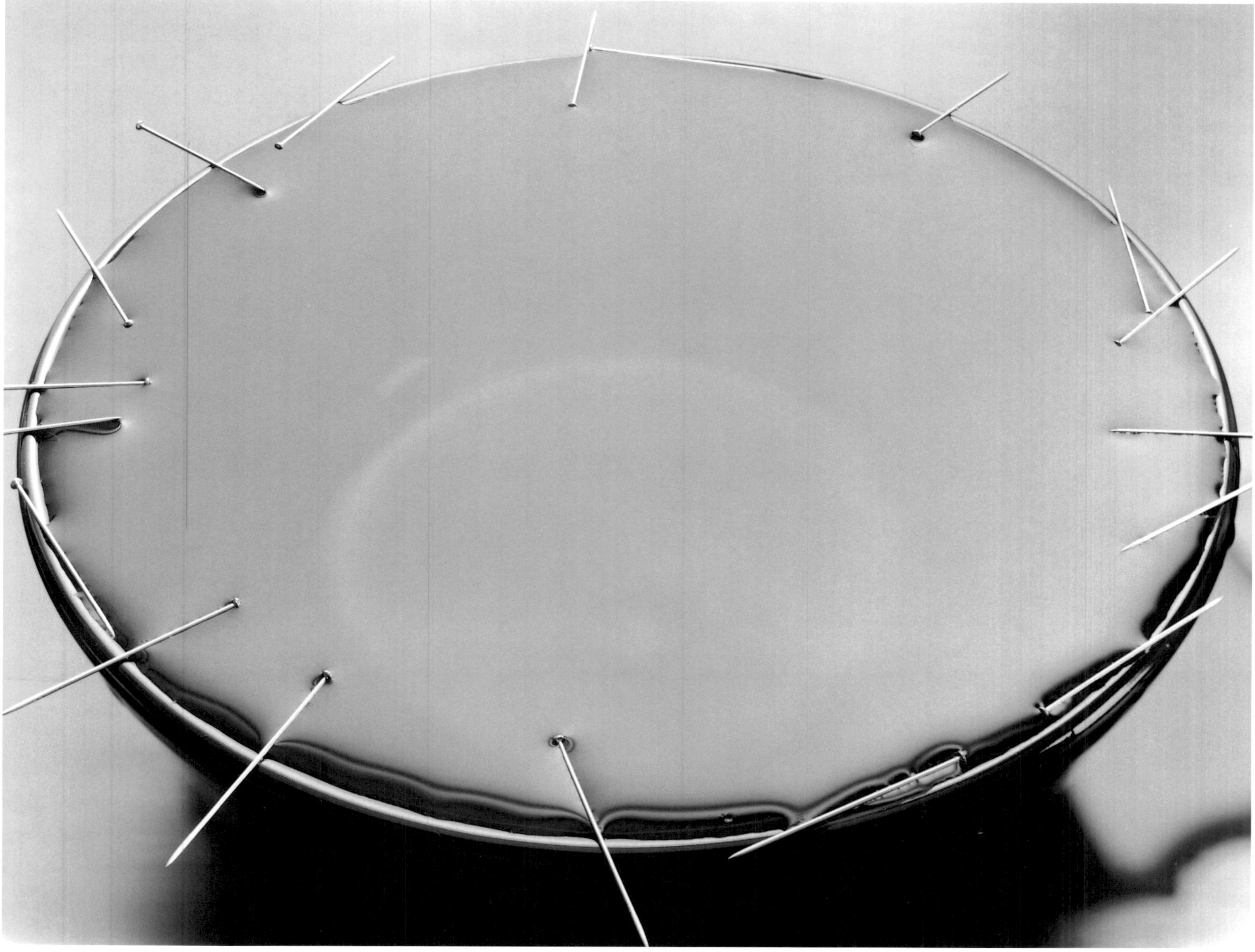

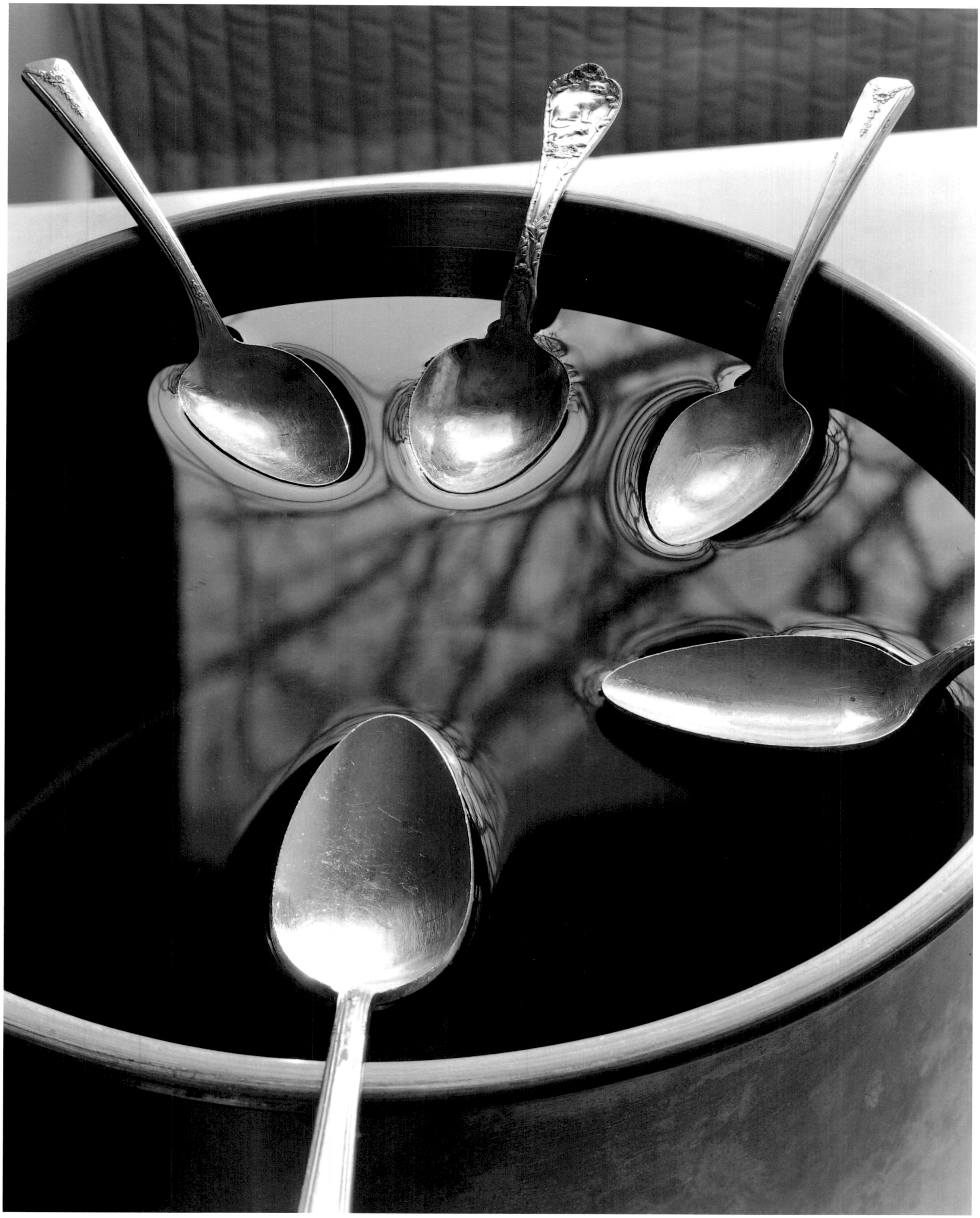

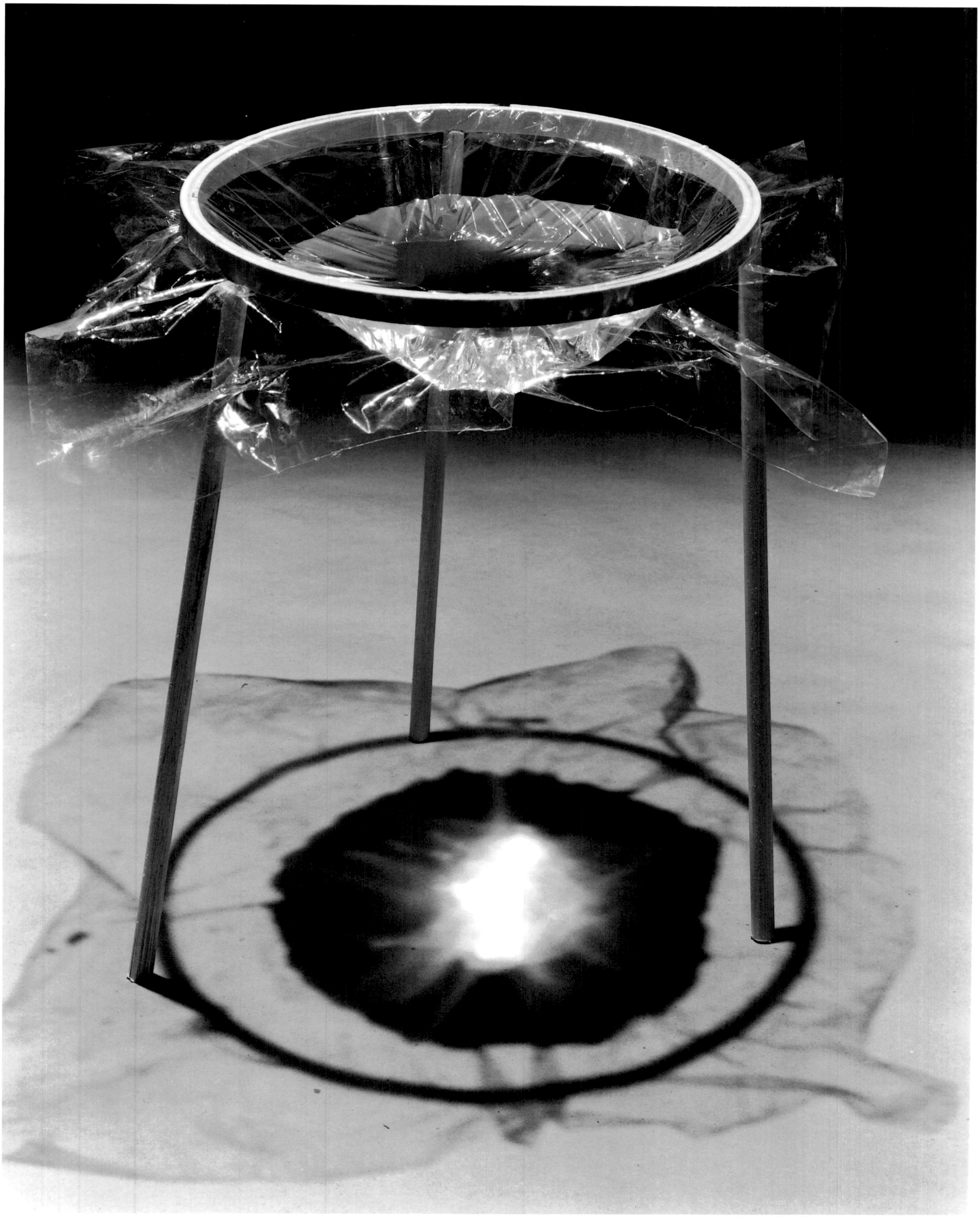

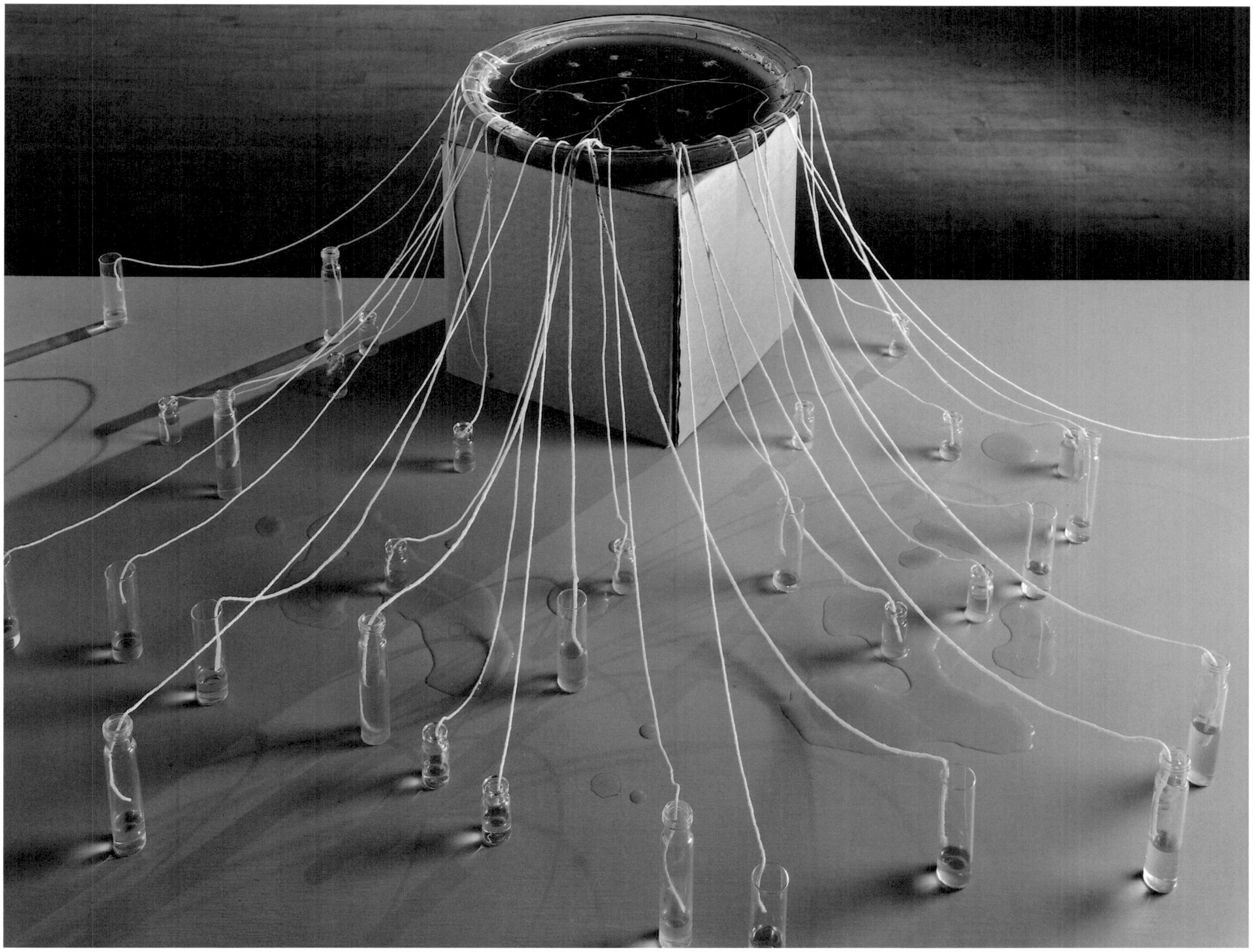

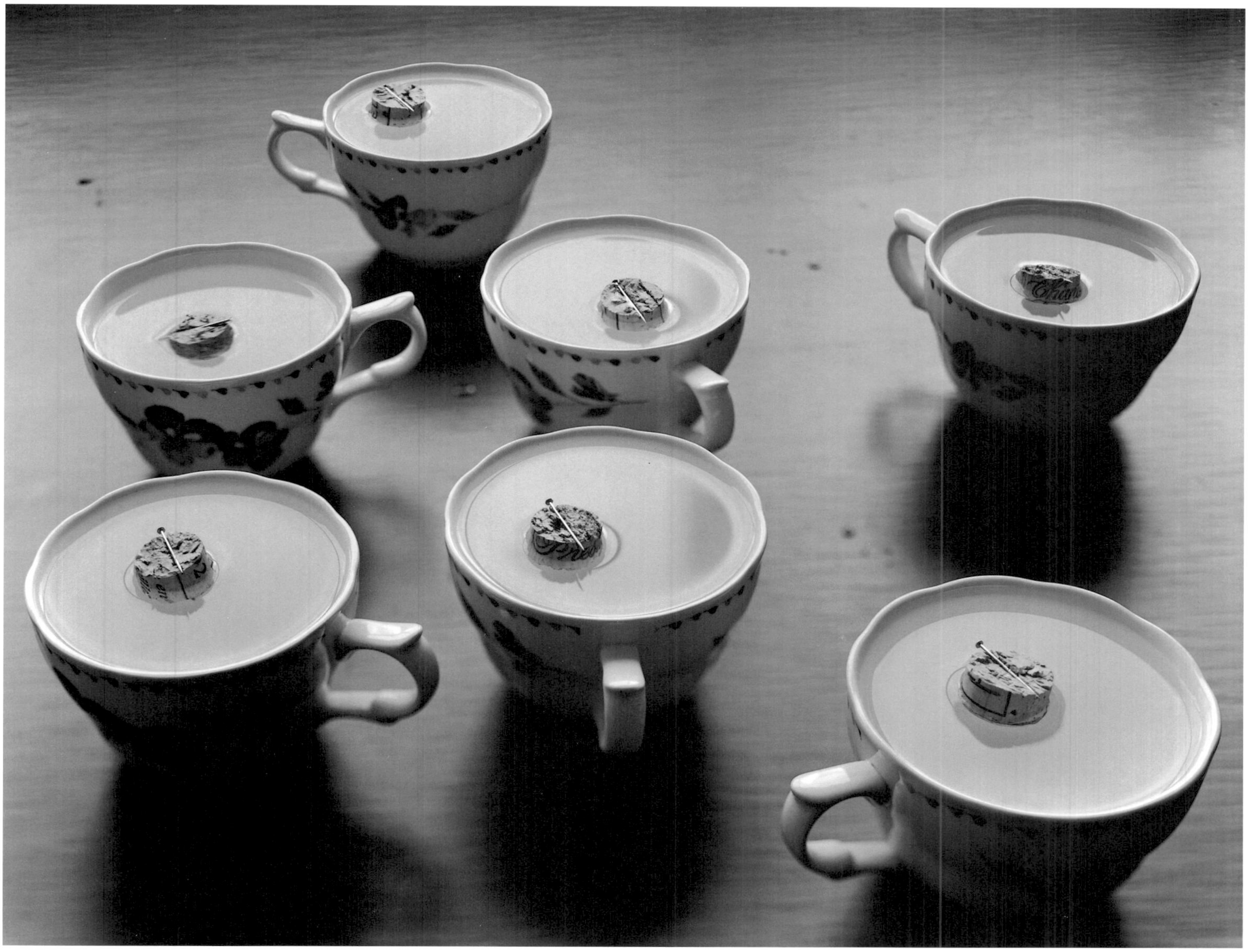

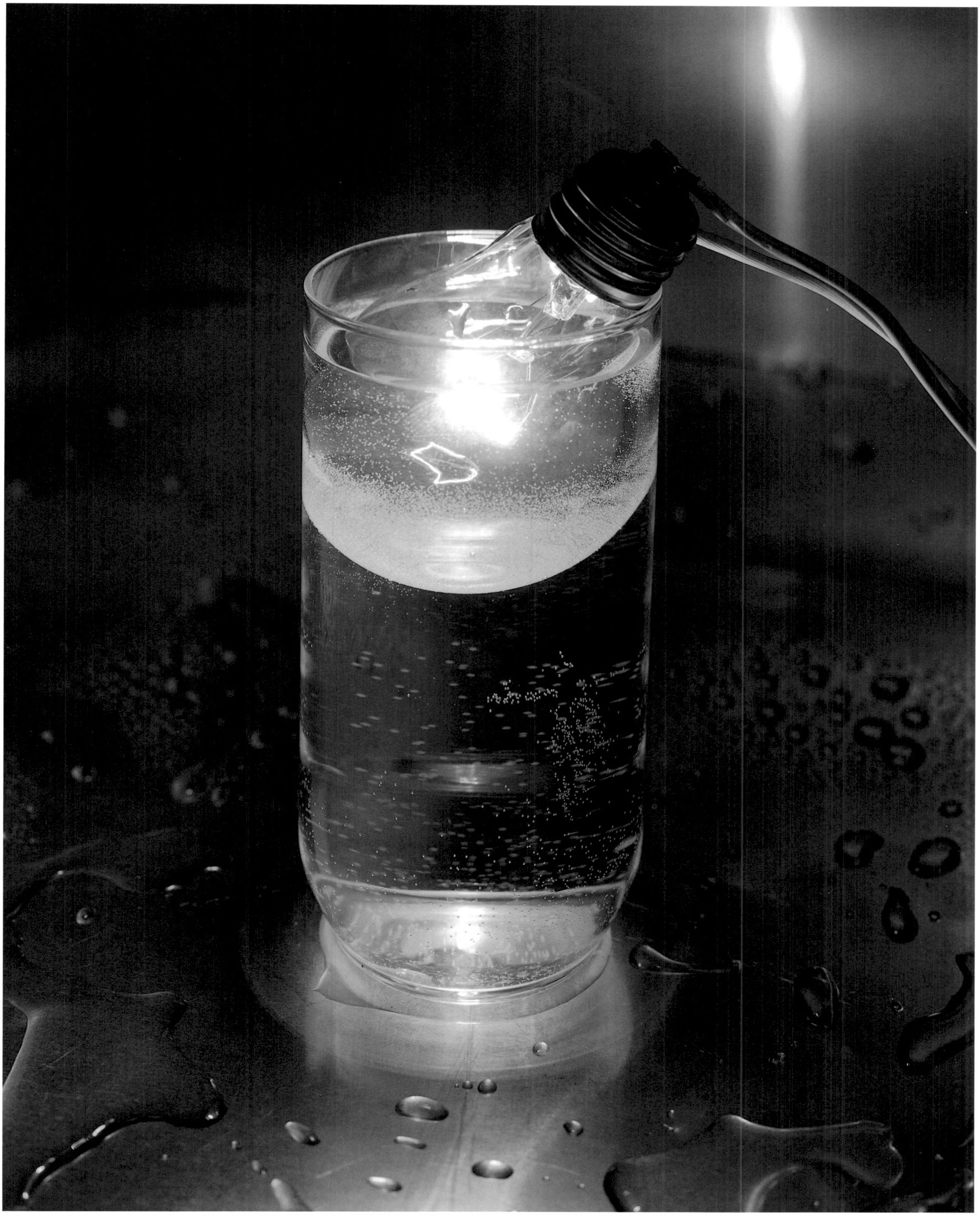

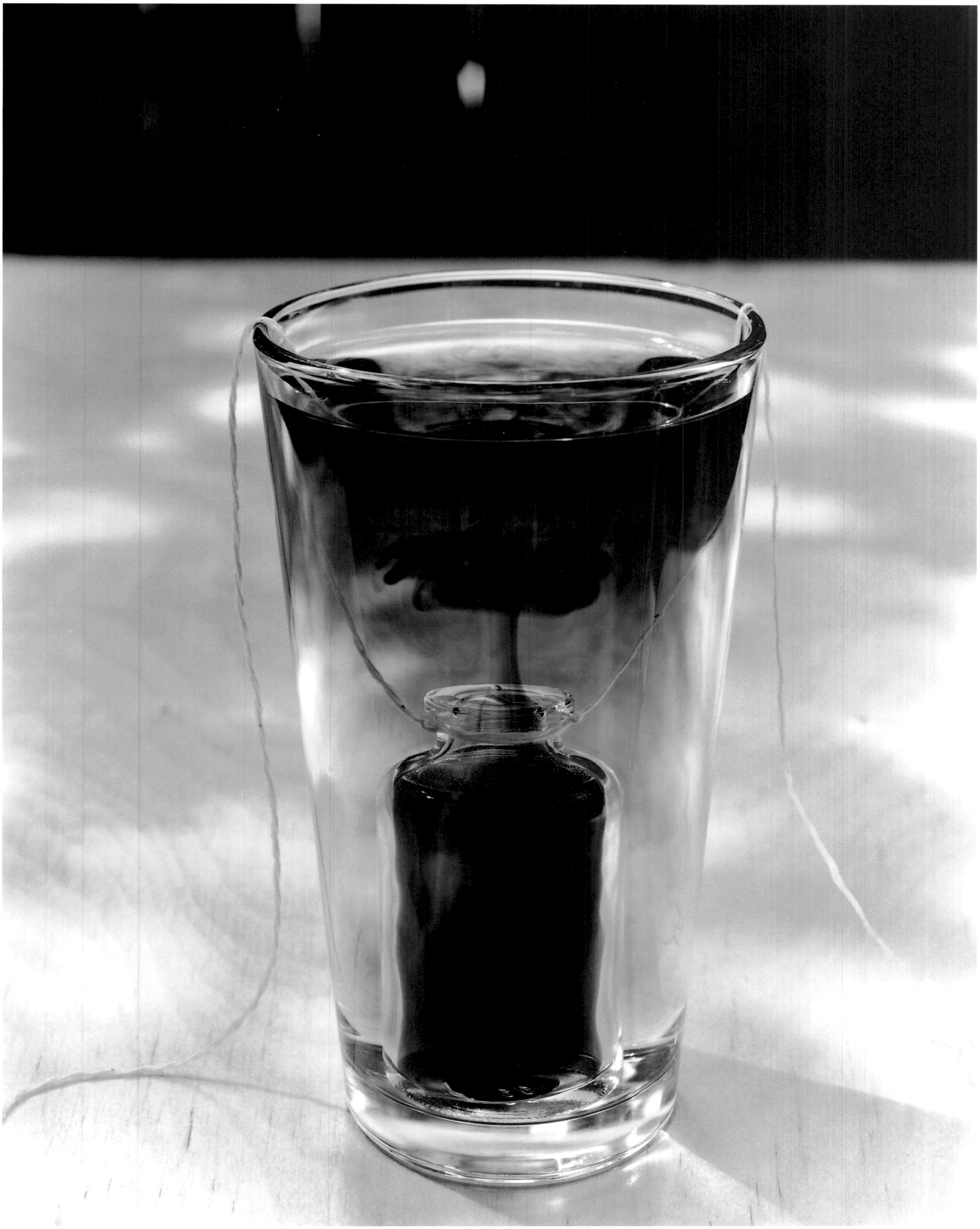

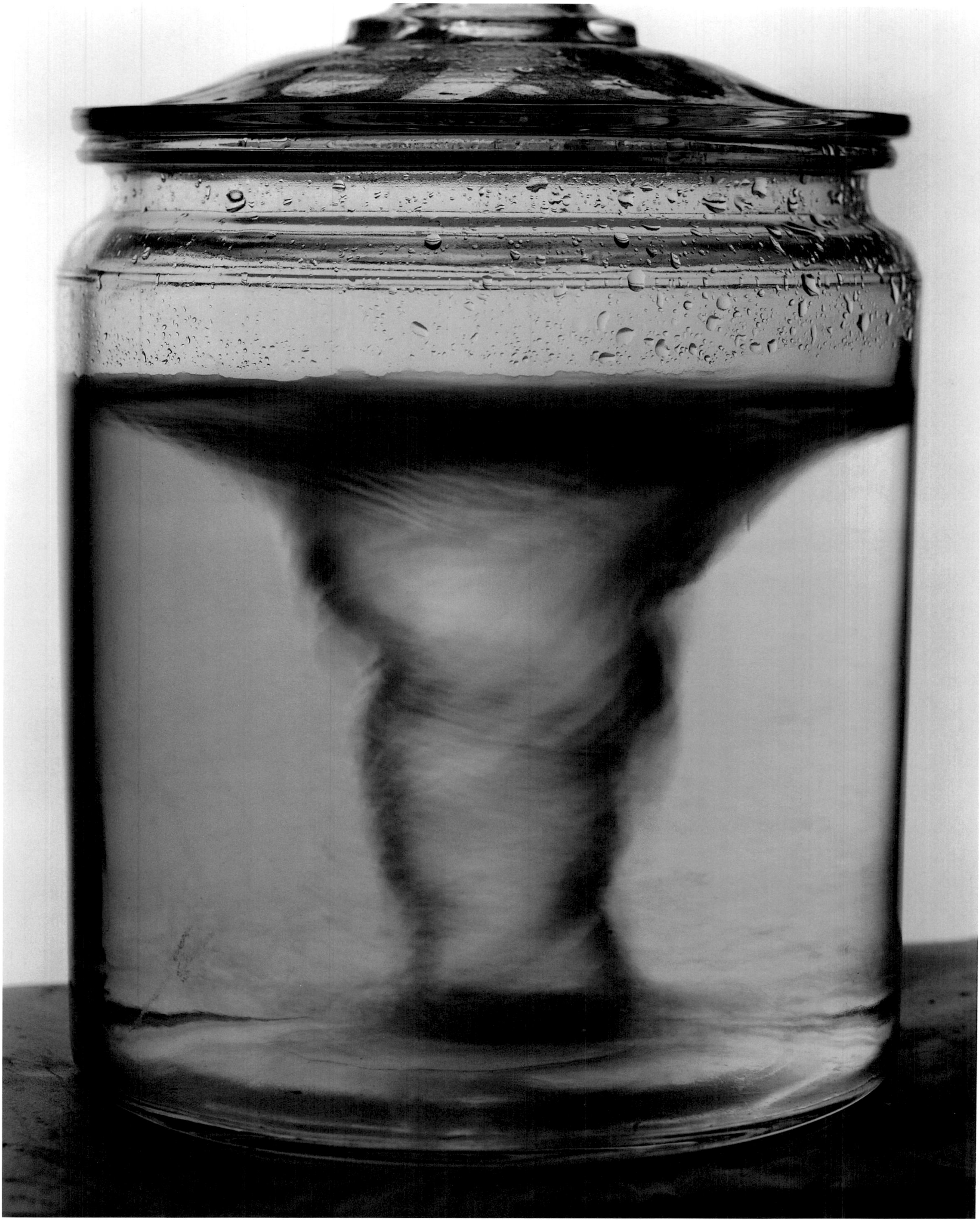

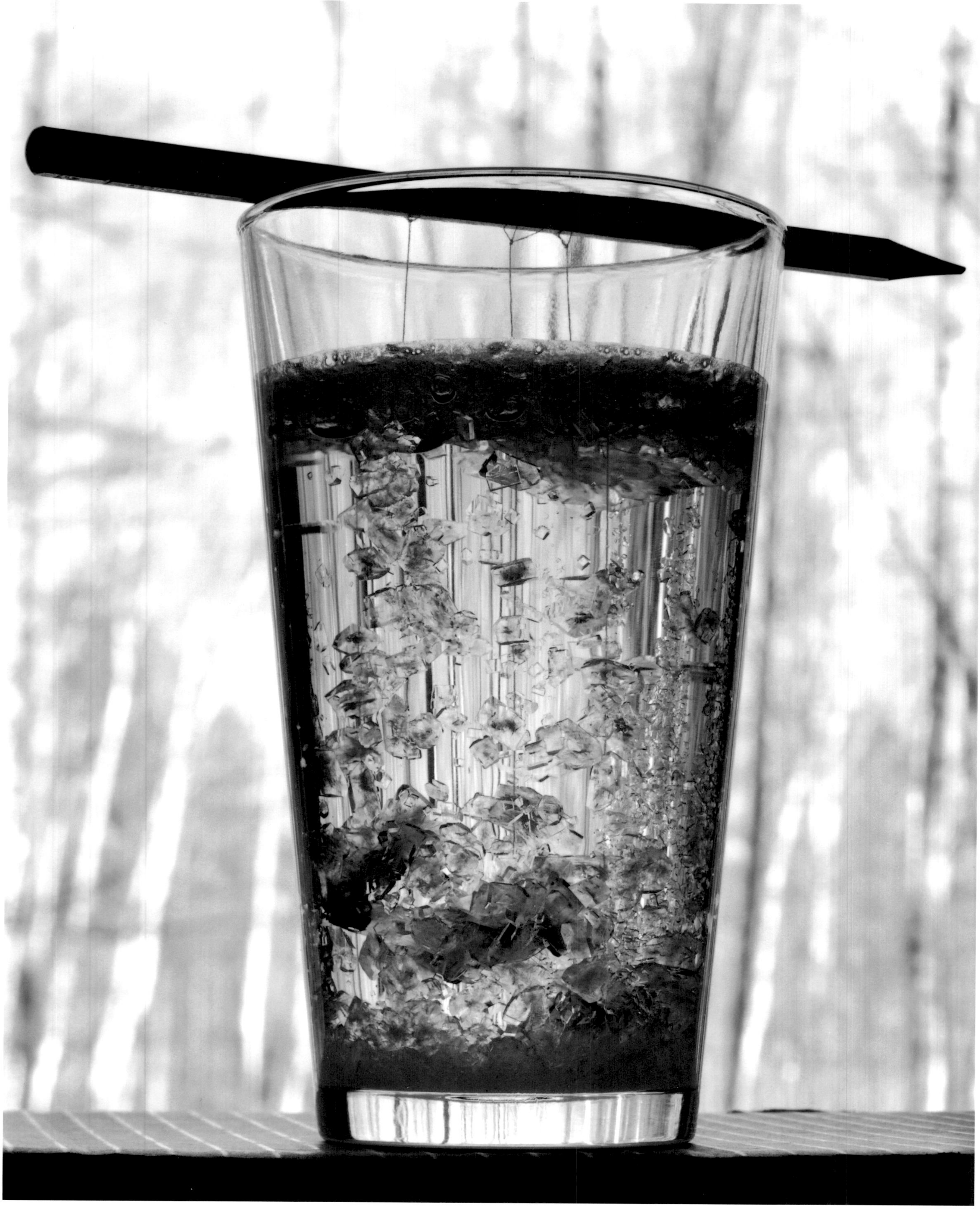

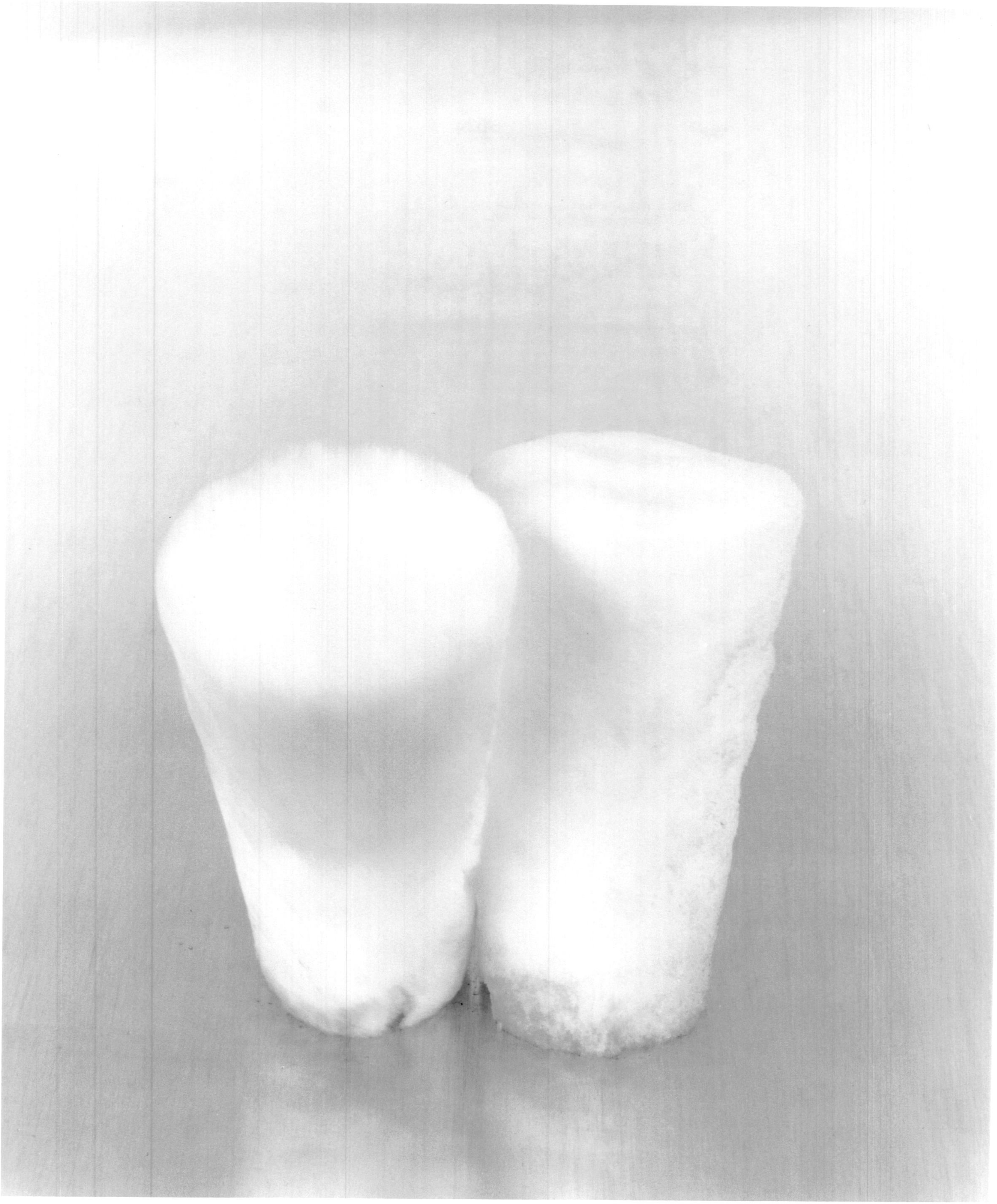

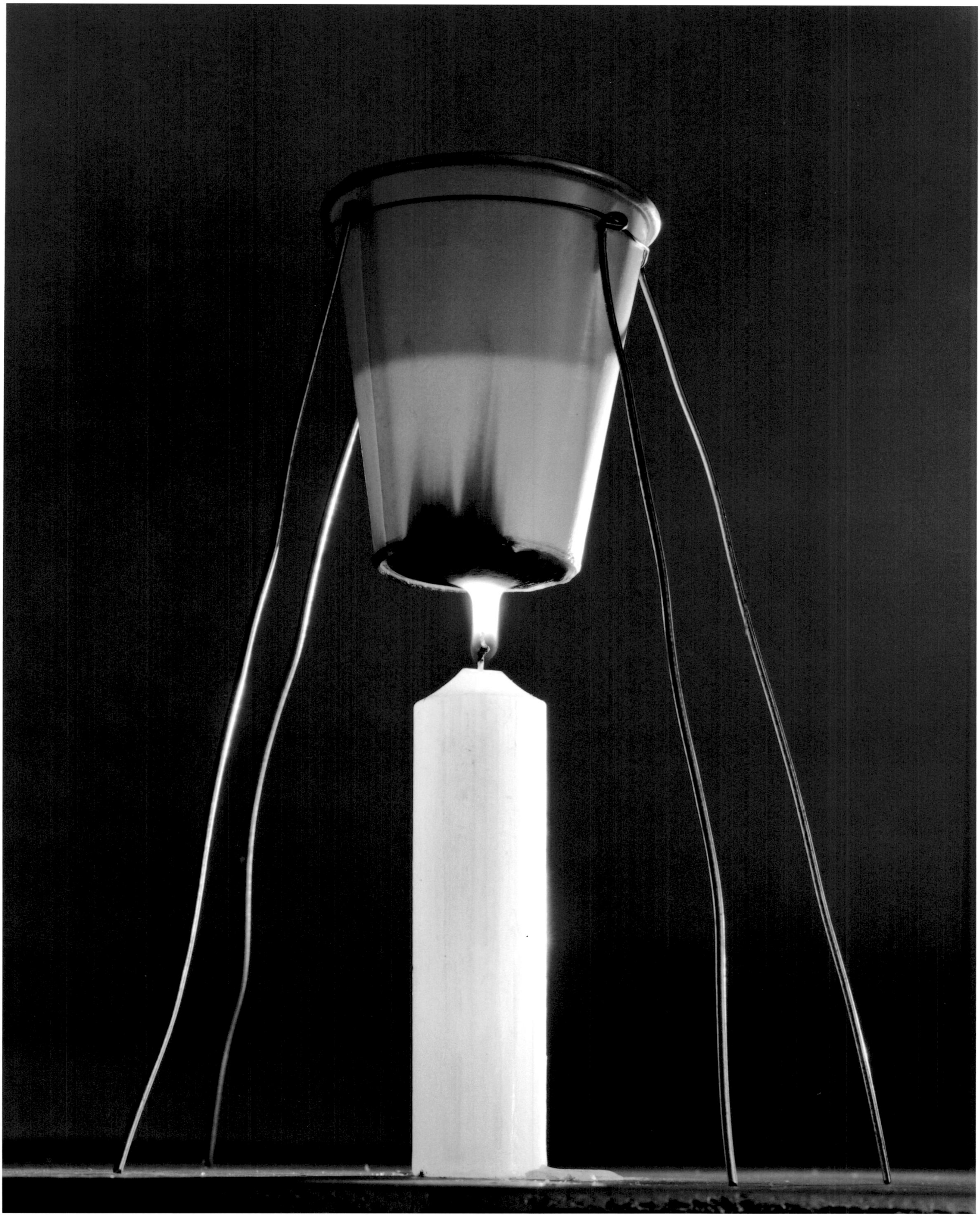

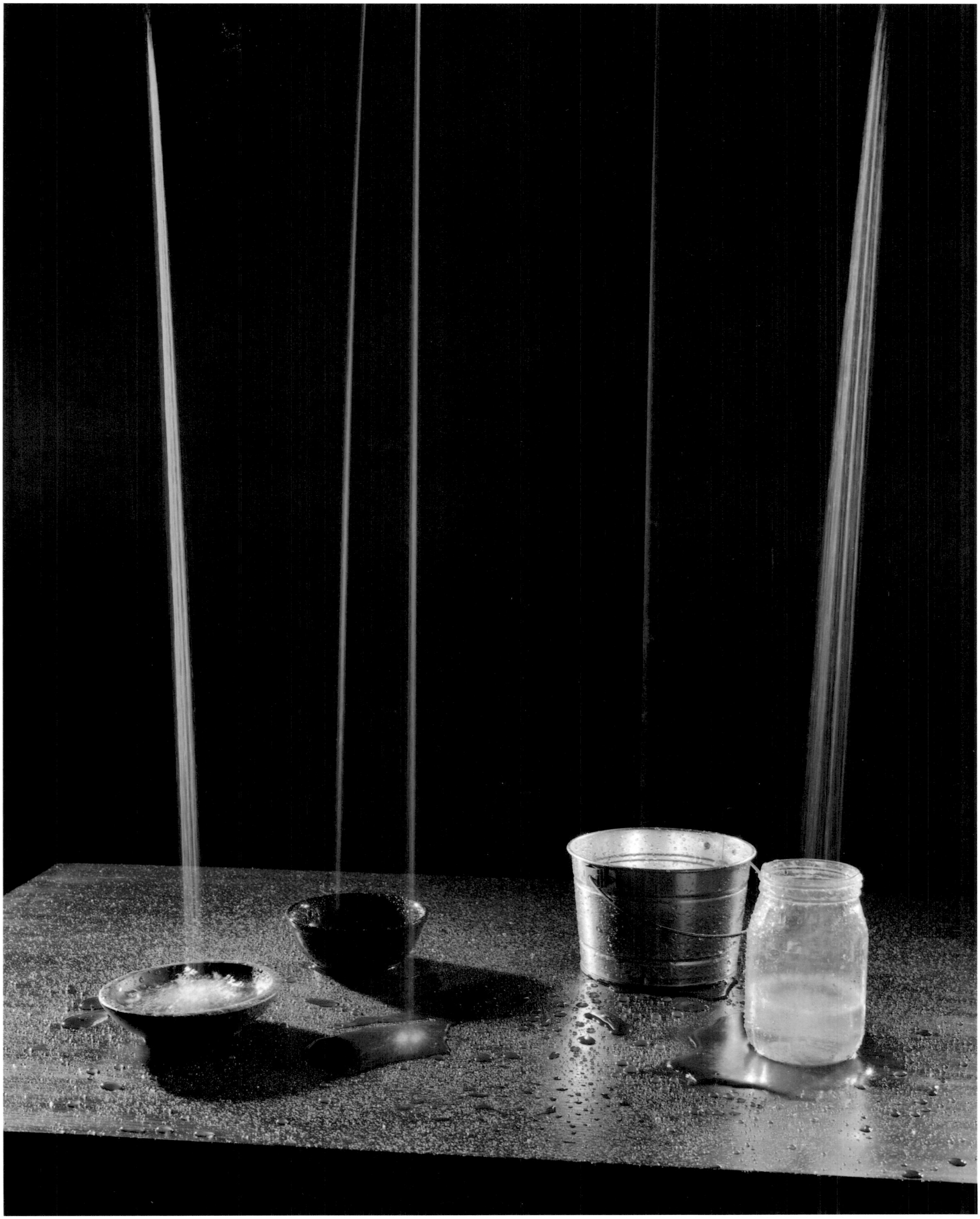

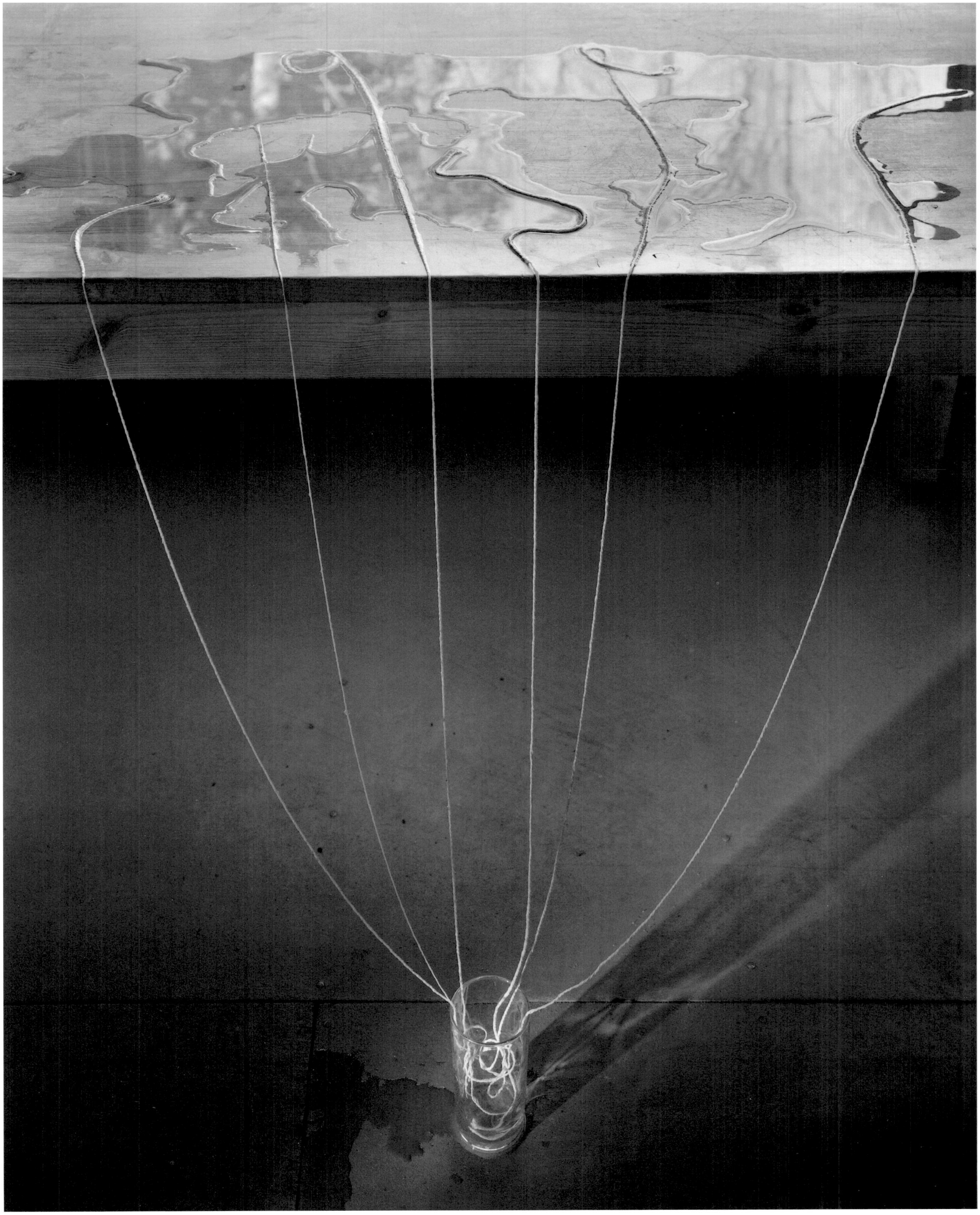

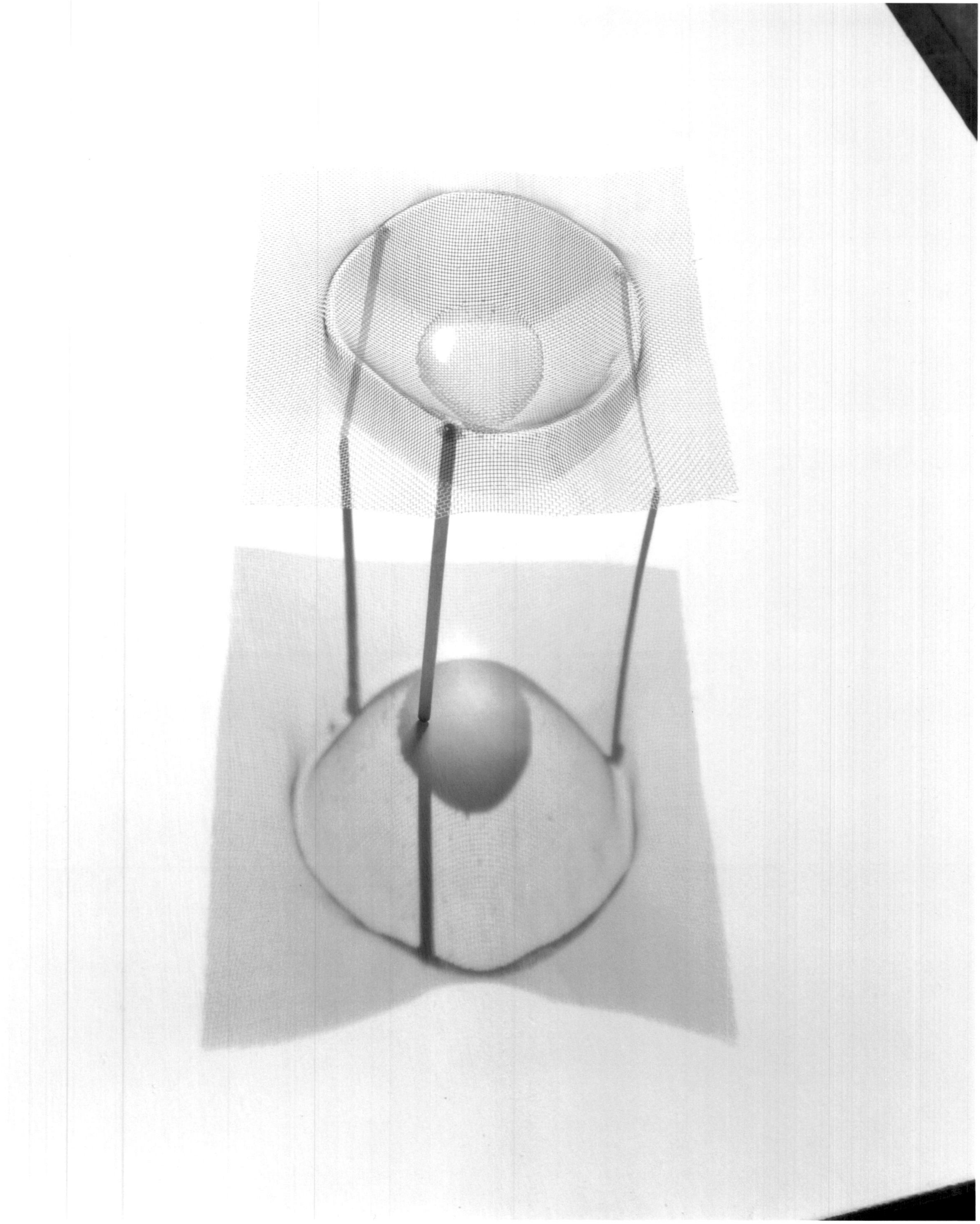

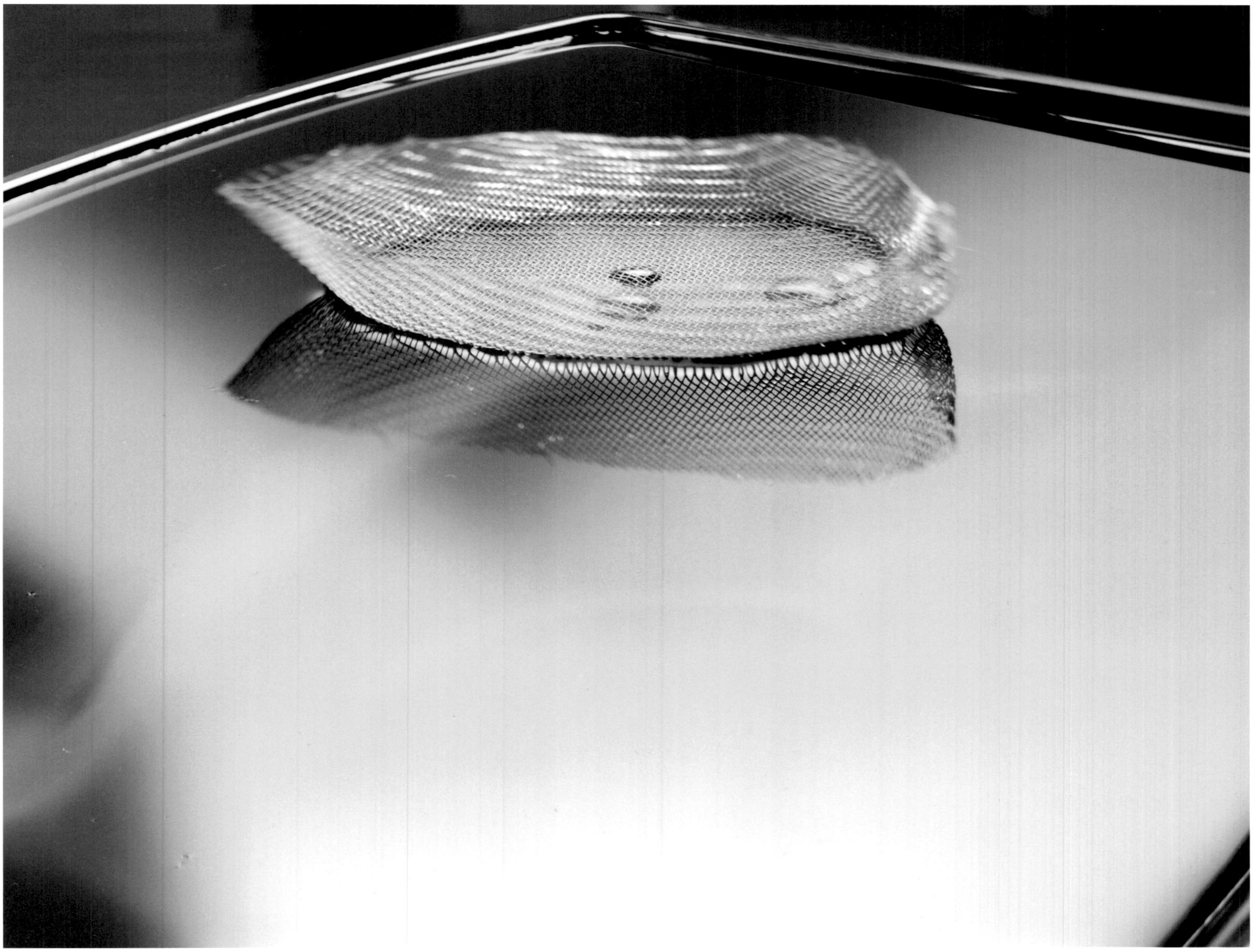

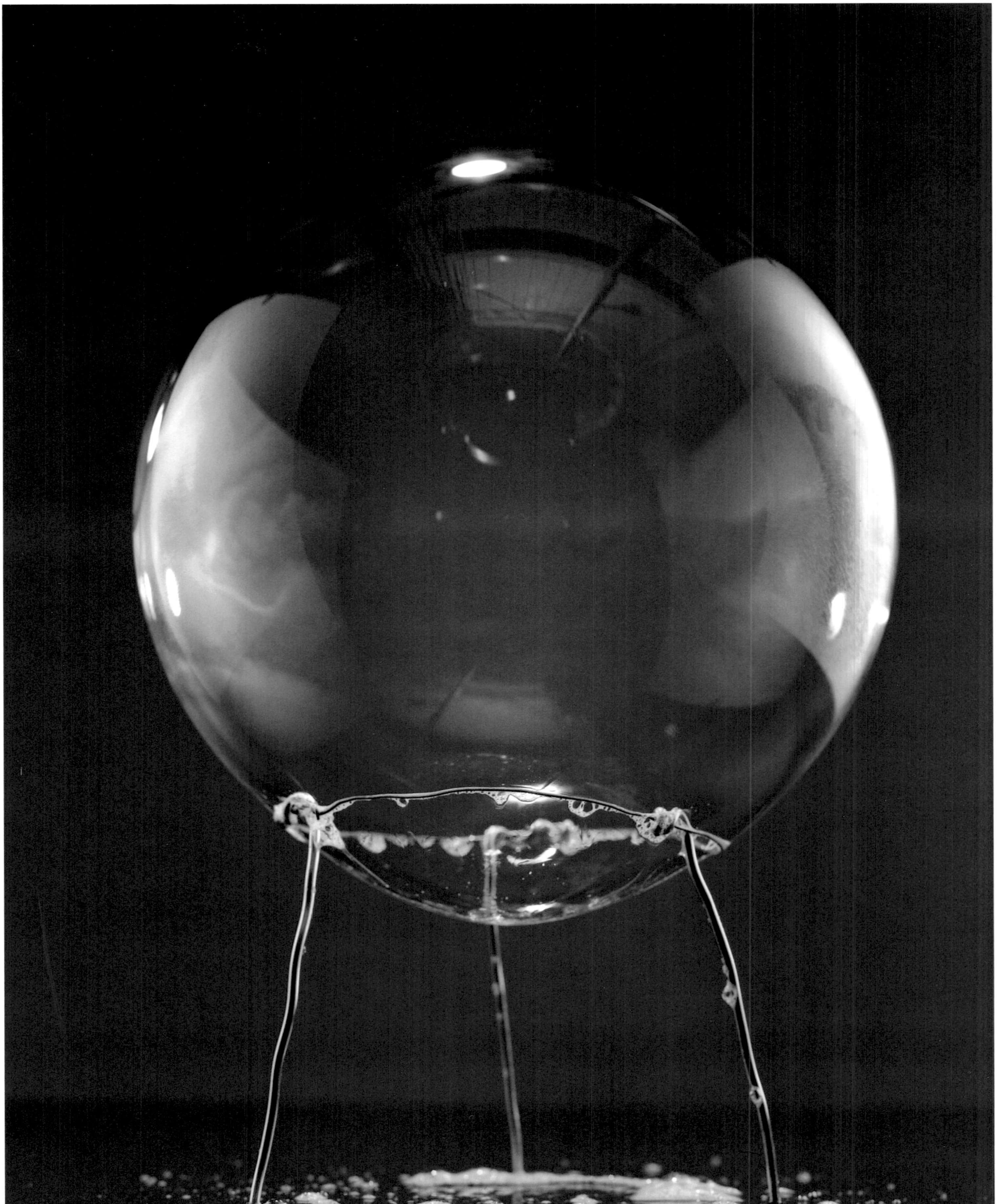